Decision Making for Regulating Chemicals in the Environment

A report prepared by the
COMMITTEE ON PRINCIPLES OF DECISION MAKING FOR REGULATING CHEMICALS IN THE ENVIRONMENT

Environmental Studies Board
Commission on Natural Resources
National Research Council

NATIONAL ACADEMY OF SCIENCES
Washington, D.C. 1975

NOTICE: The project which is the subject of this report was approved by the Governing Board of the National Research Council, acting in behalf of the National Academy of Sciences. Such approval reflects the Board's judgment that the project is of national importance and appropriate with respect to both the purposes and resources of the National Research Council. The members of the committee selected to undertake this project and prepare this report were chosen for recognized scholarly competence and with due consideration for the balance of disciplines appropriate to the project. Responsibility for the detailed aspects of this report rests with that committee. Each report issuing from a study committee of the National Research Council is reviewed by an independent group of qualified individuals according to procedures established and monitored by the Report Review Committee of the National Academy of Sciences. Distribution of the report is approved, by the President of the Academy, upon satisfactory completion of the review process.

At the request of and funded by the
U.S. Environmental Protection Agency
Contract No. 68-01-2262

Library of Congress Catalog Card Number 75-29655
International Standard Book Number 0-309-02401-3

Available from:
Printing and Publishing Office, National Academy of Sciences
2101 Constitution Avenue, N.W., Washington, D.C. 20418

COMMITTEE ON PRINCIPLES OF DECISION MAKING FOR REGULATING CHEMICALS IN THE ENVIRONMENT

J. CLARENCE DAVIES, III, Resources for the Future, Washington, D.C., *Chairman.*

RICHARD FAIRBANKS, Ruckelshaus, Beveridge, Fairbanks & Diamond, Washington, D.C.

A. MYRICK FREEMAN, III, Bowdoin College, Brunswick, Maine.

JAMES L. GODDARD, Ormont Drug and Chemical Company, Englewood, New Jersey.

JAMES D. HEAD, Dow Chemical Company, Midland, Michigan.

DAVID L. JACKSON, The Johns Hopkins University Hospital, Baltimore, Maryland.

STANLEY LEBERGOTT, Wesleyan University, Middletown, Connecticut.

NORTON NELSON, New York University, New York, New York.

GLENN PAULSON, New Jersey Department of Environmental Protection, Trenton, New Jersey.

DAVID RALL, National Institute of Environmental Health Sciences, Research Triangle Park, North Carolina.

SHELDON W. SAMUELS, Industrial Union Department of AFL–CIO, Washington, D.C.

JESSE L. STEINFELD, University of California, Irvine, California.

WILLIAM A. THOMAS, American Bar Foundation, Chicago, Illinois.

STAFF

RALPH C. WANDS, Staff Officer, Committee on Toxicology.

CHARLES R. MALONE, Staff Officer, Environmental Studies Board.

ADELE L. KING, Staff Associate, Environmental Studies Board.

Preface

The study culminating in this report was undertaken at the request of the U.S. Environmental Protection Agency, which asked the National Academy of Sciences to examine principles of decision making for regulating chemicals in the environment. The intent of the study was to assess the state of the art of this type of decision making, to identify inadequacies in current methods, and to make recommendations that could aid EPA and other regulatory agencies in making more equitable and scientifically sound decisions for controlling chemicals in the environment. The study was also intended to find ways to increase the information base upon which regulatory decisions are made and to develop methods of displaying the available information to make it more useful to the decision maker.

Within the National Research Council, responsibility for the study was assigned to the Environmental Studies Board of the Commission on Natural Resources, with advisory participation by the Committee on Toxicology of the Assembly of Life Sciences. These units sponsored a symposium on September 19–20, 1974, at the National Academy of Sciences in Washington, D.C., to explore the state of the art of assessing benefits derived from chemicals that pose environmental threats, balancing costs and risks of chemicals with their benefits to society, and making appropriate regulatory decisions on the basis of such assessments. Persons attending the symposium were asked to present their views on these subjects, and the informal presentations initiated discussions that led to a preliminary work plan for the remainder of the study. That plan called for a series of case studies of past decisions by government and industry to regulate or restrict the use of chemicals. It also provided the framework for a working conference designed to furnish the principal basis for this report to EPA.

Subsequent to the symposium the NRC appointed a Committee on Principles of Decision Making for Regulating Chemicals in the Environment under the auspices of the Environmental Studies Board and the Committee on Toxicology. Several case studies were solicited for the Committee's use: these included a paper on EPA's decision to establish residue tolerances for hexachlorobenzene, three papers with different views on EPA's decision to cancel the registration of DDT, two papers on decisions to restrict the use of polychlorinated biphenyls, and a paper on an industry's decision not to market certain uses of polybrominated biphenyls. The case studies were drawn upon by the Committee in preparing its report. They are not published here but are available in limited number upon request from the NRC's Environmental Studies Board at 2101 Constitution Ave., N.W., Washington, D.C. 20418.

The Committee also had available to it some preliminary work on a benefit–cost analysis of lead in the environment prepared under contract for EPA by the CONSAD Research Corporation, Pittsburgh, Pennsylvania. We are grateful to the CONSAD Corporation and to Dr. Fred H. Abel of EPA for making their early analyses available to us.

In pursuing the study plan developed at the symposium, the Committee organized a working conference on regulatory decision making which was held in New Orleans, Louisiana, February 19–23, 1975. Approximately 50 participants and observers (Appendix A) were invited by the Committee to consider various aspects of the decision-making process. The participants, after hearing an address by the head of EPA's Office of Toxic Substances (Appendix B), were assigned to one of eight panels charged with writing a working paper for use by the Committee in preparing its report to EPA. Although general subject matter assignments were made to each panel, no attempt was made to prevent coverage of similar subjects by more than one panel or to resolve conflicting views as they developed among the panels during the conference. The Committee encouraged free and open expression of all viewpoints so as to have the widest possible range of views to draw upon in preparing its report.

The working papers prepared at New Orleans provided the principal body of information upon which our report was developed. They are included here (Appendixes C–J) so that the reader can see which views were accepted, rejected, or modified by the Committee.

The final input to our report was provided by a public meeting held by the Committee on March 5, 1975, at the National Academy of Sciences. About 90 persons attended that session and heard the preliminary views of the Committee on regulatory decision making. The attendees were encouraged to express their views, which were transcribed and studied by the Committee before our report was prepared. The transcription of the proceedings of the public meeting is not included in this report, but it is also available in limited number upon request from the Environmental Studies Board.

The reader should note that of the material included in this volume only the report and recommendations of the Committee have been critically reviewed by the NAS. Views expressed in the appendixes are those only of the persons who contributed to the working papers. Those papers are not formal portions or chapters of our report.

Finally, this report can be viewed as a companion volume to the recent 1975 NAS publication, *Principles for Evaluating Chemicals in the Environment.* That publication is a technical and scientific assessment of methods for evaluating the risks and hazards associated with chemicals that pose potential environmental threats. Our report refers frequently to this earlier volume in the discussions of risks and hazards and how these can be compared to benefits for regulatory decision making and the effective presentation of information to the decision maker.

J. CLARENCE DAVIES, III, *Chairman*
Committee on Principles of Decision Making
for Regulating Chemicals in the Environment

Contents

APPENDIXES

Summary of Conclusions and Recommendations

This report analyzes the key aspects of the process for making regulatory decisions about chemicals in the environment. It focuses primarily on industrial chemicals and on decision making at the federal level. The factors external to a regulatory agency—the statutory framework and the participation of groups outside an agency—are analyzed first. The use of information within the agency is then considered.

The Committee's conclusions and recommendations deal with four areas that we consider to be the most important for an equitable and rational decision-making process. These four areas are: (1) the statutory and organizational basis for regulation; (2) openness and access to the decision-making process; (3) the availability of adequate and reliable information; and (4) the proper use of analysis.

The statutory framework for regulating chemicals is a major determinant of both what will be regulated and how regulatory decisions will be made. Existing laws contain a number of inconsistencies on such matters as who should bear the burden of proof, how much discretion shall be left to the administrative agency, the weight to be given to health as against other factors, and the nature and extent of judicial review of decisions. There are general principles that we believe can and should be applied to reconcile some of these inconsistencies. In addition, the statutory framework for regulating chemicals could be improved if there were a more systematic and comprehensive basis for collecting information about industrial chemicals and for regulating chemicals that pose an unreasonable hazard.

The decision-making process should be as open as possible to outside participation. An equitable decision process is one where the consideration given to the interests of potentially affected individuals is proportional to the anticipated effects of the decision on those individuals. In other words, the decision process should include the groups in society who will feel the impact of the decision, and it should reflect the degree of impact of the decision on those groups. Openness in decision making will improve the quality of the information used to make the decision and will facilitate implementation and acceptance of the decision.

It is clear that adequate and reliable information is fundamental to making sound decisions. The quantity and quality of *all* types of information used for regulating chemicals need to be greatly improved. The information available to the decision maker will never be perfect and is often not satisfactory, but a number of steps can be taken to provide more adequate information.

The regulatory agency should be as explicit as possible about the factors that enter into the decision, both during and after the decision-making process. The hazards of the chemical under consideration, the bureaucratic costs of regulation, the expenses to business of complying with the regulation, the dislocation of economic resources, the availability of substitute products, the potential benefits foregone when scientific research is diverted or discouraged, and other relevant factors should be clearly displayed for the decision maker and for the public.

There is no scientific formula for making regulatory decisions. Decisions about regulating chemicals in the environment always involve values about which the affected parties disagree; thus the values of the decision maker will play a crucial role in the outcome. There is no satisfactory way to summarize all the costs or benefits of regulatory options in dollars or other terms which can be mathematically added, subtracted, or compared. In short, there is no substitute for an experienced decision maker exercising good judgment. However, the techniques developed by decision theory and benefit–cost analysis can provide the decision maker with a useful framework and language for describing and discussing trade-offs, noncommensurability, and uncertainty. They also can help to clarify the existence of alternatives, decision points, gaps in information, and value judgments concerning trade-offs.

The four areas covered by our conclusions and recommendations are closely interrelated. The statutory basis of regulation will determine in part what information is made available to the decision maker and how open the decision-making process is. The use of analysis within the agency will depend on the adequacy of information, the openness of the process, and the language of the relevant laws.

The initial decision process for which a regulatory agency is responsible should be so strengthened by the proper use of adequate scientific and economic information and so broadly based that there will be less likelihood that the ultimate issue will be decided in the courts. The Committee believes that the recommendations below will help to achieve this goal.

(For a better understanding of the recommendations the reader is encouraged to read the text surrounding each recommendation. To facilitate this the page on which each recommendation occurs is given.)

THE STATUTORY AND ORGANIZATIONAL BASIS FOR REGULATION

1. As a general principle the burden of proof that society will obtain a benefit from a *new* use of a chemical should rest with those proposing such use. It may be desirable to make statutory changes to reflect this principle (page 18).

2. Once the government has made a reasonable case that the challenged use of an *existing* chemical creates an excessive hazard to human health or to the environment, the burden of producing evidence should shift to the proponent of use, who must then make an appropriate showing that the continued use is desirable (page 18).

3. Statutory provisions should not preclude consideration of any relevant factors in the decision-making process. Those provisions that prevent such consideration should be considered for possible amendment (page 20).

4. Whenever possible, and without precluding administrative consideration of relevant factors, Congress should provide increased and consistent statutory guidance as to the relative importance that should be given to health, environmental, and economic factors in regulating chemicals (page 20).

5. EPA should undertake a study to (1) identify neglected areas in hazard or pollution control, and (2) determine whether existing legal authorities are available and unused or whether new laws should be sought from Congress. Examples of such needs include the problem of choosing optimal waste disposal methods considering the total environment, and the question of indoor air pollution (page 17).

6. Congress should review the adversary procedures that have led to unduly protracted hearings (such as pesticide cancellation) and deter-

mine how best such marathon decision making can be shortened (page 30).

7. Exposure to many chemicals comes simultaneously from two or more media. An interagency committee should be established to ensure that each agency's standards reflect appreciation of such multiple exposures (page 31).

8. Within the Committee there were divergent opinions on the optimum organizational structure for making chemical regulatory decisions. It is recommended that Congress give consideration to alternative decision-making structures. In particular, it should consider whether responsibility should be vested in a board or commission with fixed terms in office or in a single administrator (page 22).

OPENNESS AND ACCESS TO THE DECISION-MAKING PROCESS

9. The essential elements of decision making should be part of the public record. The agency should publish a "white paper" for each important regulatory action undertaken. The paper should include the key details of the economic, legal, scientific, and other considerations taken into account in reaching the decision. It should be issued when the agency decides to take some action but sufficiently in advance of a final decision to permit considered response. An important decision to take no regulatory action, or to defer such action, should also be accompanied by a "white paper" (page 28).

10. Any information available to an agency on the hazards of a chemical that is regulated by that agency should not be considered proprietary and should be available for public inspection in a timely fashion during and after the decision-making process (page 28).

11. The early and open exchange of information and opinions on a proposed decision should be encouraged to reduce the current dependence on subsequent judicial challenge. The EPA Administrator should hold public hearings at the earliest feasible stages of the decision process. He also should facilitate prehearing exchange of information among parties (for example, through depositions, interrogatories, and other discovery procedures) (page 21).

12. At appropriate points in the decision-making process, the agency should actively seek to identify the affected parties and solicit suggestions and comments from them. Ways should be explored to better represent the interests of future generations (page 29).

13. The press as well as other interested parties should be told when a standard-setting process begins and when the proposed standard is ready

for publication; further discussions during this process should occur as often as warranted (page 27).

14. EPA and other agencies should initiate programs aimed at training and encouraging citizens to participate in the decision-making process (page 29).

15. The Department of Commerce should develop an educational resource to help small businesses acquire the information on chemical regulatory matters that is at present routinely available to large corporations and major trade associations (page 26).

16. All *ex parte* communications, including those from Congress, members of the Executive Branch, private corporations, and citizen groups, on any adjudicatory decision pending before a regulatory agency should be made public with sufficient time for comment before a decision is made (page 28).

THE AVAILABILITY OF ADEQUATE AND RELIABLE INFORMATION

17. The quality of chemical regulatory decisions is dependent largely upon the adequacy of the available information. To develop an adequate data base, research efforts in basic clinical and environmental toxicology and epidemiology and in economic analysis must be strengthened, and professional training in these areas must be supported. An interagency committee consisting of the relevant federal research and regulatory agencies should be established to maximize the use of existing information (page 53).

18. Existing toxicological and other information systems related to the regulation of chemicals should be examined with a view to improving their coordination and use (page 13).

19. For optimal regulatory decision making, a procedure to conduct retrospective analyses of the impact of given decisions should be adopted. This should emphasize evaluation of the accuracy of the predictive models for health, economics, and environmental sciences in the original decision-making process. EPA should also develop a formal system that reviews and identifies information necessary for optimal future decision making and implements the appropriate programs to generate this information in a timely fashion (page 37).

20. All agencies that regulate chemicals should establish procedures for external scientific review of the technical data base presented to the decision maker. The results of the review should be available directly to the decision maker and to the public as early as feasible. Specifically, the

major research and development programs undertaken by EPA should be reviewed routinely by panels of qualified experts (page 24).

21. EPA should make greater use of committees of experts who represent the spectrum of potential viewpoints, particularly in helping to anticipate future chemical control problems (page 33).

22. EPA's Science Advisory Board should be permitted to formally request the Administrator to subpoena information needed for the Board's consideration and to allow the Board to investigate new areas. The request and the Administrator's response should be a matter of public record. The existence and responsibilities of the Science Advisory Board should be mandated by legislation (page 25).

23. The power to subpoena expert witnesses, now available to EPA under the Federal Environmental Pesticide Control Act, should be extended to proceedings under other statutes to develop the fullest and fairest public record (page 25).

24. To improve the quality of scientific evidence, the right of counsel to cross-examine expert witnesses should be guaranteed (page 25).

25. Other federal agencies should have the opportunity to participate in internal EPA deliberations early in the consideration of regulatory options. The same opportunity should be given to EPA by other agencies that regulate chemicals (page 30).

26. Federal agencies regulating chemicals should establish a formal and regular method for obtaining information from state agencies (page 27).

27. The Department of Health, Education, and Welfare, in conjunction with EPA, should attempt to develop a hazard rating system, placing particular emphasis on evaluation of use patterns (page 52).

28. The Department of Health, Education, and Welfare and EPA should establish a task force including, among others, representatives of the chemical industry and the scientific community to develop a system for making unpublished or proprietary data about chemicals available to governmental agencies (page 53),

29. EPA in cooperation with the Department of Health, Education, and Welfare should develop and use monitoring systems that can detect changing patterns in concentrations of specific toxic substances in biological tissues. They should also develop and use population surveillance systems that reflect changes in illness and death patterns due to environmental pollutant exposure. Data from monitoring systems and from other sources should be used to adjust past decisions when necessary (page 54).

THE PROPER USE OF ANALYSIS

30. Highly formalized methods of benefit–cost analysis can seldom be used for making decisions about regulating chemicals in the environment. Thus the development of such methods should not have high priority. However, the benefit–cost and decision frameworks described in this report can be useful in organizing and summarizing relevant data on regulatory alternatives which the decision maker must review (page 44).

31. Value judgments about noncommensurate factors in a decision such as life, health, aesthetics, and equity should be explicitly dealt with by the politically responsible decision makers and not hidden in purportedly objective data and analysis (page 50).

32. The decision process should require the agency's technical staff to present a full set of options with a corresponding range of cost–benefit–hazard data and explicit statements on the confidence limits of each analysis (page 33).

33. EPA should adopt, whenever scientifically possible, a generic approach, as opposed to an ad hoc procedure, for the regulation of chemicals (page 33).

34. EPA should ask the American Bar Association to cooperate in a study of the way health and environmental information and economic analyses should be introduced and considered in judicial proceedings (page 22).

1 Introduction

During the past few years, our society has been faced with a large number of regulatory decisions about specific chemical compounds in the environment. The federal government has had to make difficult choices about chemicals such as hexachlorobenzene, PCBs, benzidine, mercury, vinyl chloride, and DDT. The belief that the decision-making process for regulating chemicals could be improved prompted the U.S. Environmental Protection Agency (EPA) to contract with the National Academy of Sciences (NAS) for a study to analyze the process, devise improvements, and formulate general principles applicable to making regulatory decisions for chemicals in the environment.

A more immediate factor leading to the commissioning of this study was EPA's anticipation of passage of some version of the "Toxic Substances Control Act," a measure which has been debated annually by the Congress since 1971. The proposed toxic substances bill is a broad measure with little precedent, and EPA has taken steps to plan ahead for the bill's enactment. In 1972, it enlisted the aid of the NAS to determine the appropriate methods or protocols for evaluating the hazards of chemicals in the environment. The report of that study, chaired by Dr. Norton Nelson, has recently been published (NAS 1975b). It comprehensively examines methods for evaluating the hazards of chemicals to human health and the environment. The present study logically follows the earlier report by examining the entire decision-making process and focusing on how evaluation of hazards and other information about chemicals in the environment are to be incorporated in regulatory decision making.

At the request of Congress, the NAS has also initiated a larger study of decision making in EPA. The study, budgeted at $5,000,000, will be completed in mid-1977. This larger study should provide a good opportunity to expand on the conclusions of the present study, to review their validity, and to cover areas which the limited resources available for this study did not permit us to explore.

The chemicals considered in this study are primarily industrial organic chemicals. However, the principles developed and the recommendations made should also be generally applicable to the regulation of gross pollutants (such as carbon monoxide or sulfur oxides) and of other types of chemicals. Drugs, food additives, and radiation hazards have been excluded primarily because the regulatory structure applicable to them is substantially different from that applicable to the chemicals considered in this report, and partly because the time and resources available to this committee were limited.

The portions of the report dealing with the analysis of information for making decisions are structured around the concepts of costs and benefits. As will be made clear, this should not be taken to mean that we believe that traditional benefit–cost analysis should be used to make chemical regulatory decisions. Benefit–cost analysis, as used in this report, means simply "an evaluation of all the benefits and the costs of a proposed action" (NAE 1972:2). By this definition, benefit–cost analysis is a broader concept than the traditional economic accounting of monetary values in that it encompasses all of the major positive and negative effects of a proposed action. As Ruth Mack stated, "Benefit–cost analysis, broadened to cover all significant advantages and disadvantages, is simply another name for comprehensive rational analysis. It can be structured to capitalize on the methodological contributions of economics without chaining it to the conventional subject matter of economics" (Mack 1971:76).

The ultimate test of the usefulness of this report is the extent to which it changes the behavior of those responsible for making regulatory decisions. Thus we have tried to take into account the political context, timetables, and other constraints that operate on decision makers, while still not limiting ourselves to those recommendations that can be implemented immediately or easily.

The report differs from many of those issued under the auspices of the NAS in that the subject matter inherently involves questions of values and political philosophy. There can be no single correct objective answer to the types of questions we have posed. The advantage the NAS and its

National Research Council has in dealing with such questions is the ability to bring together able and experienced people from a variety of backgrounds and disciplines to spend time in thought and discussion. The report reflects much thinking and lengthy discussion by a diverse group, and it is our hope that it will contribute to the never-ending search for better ways to make important decisions in the best interests of society.

2 Characteristics of the Regulatory Process

TYPES OF PROBLEMS

Although this study is limited to the regulation of chemicals in the environment by agencies of the federal government, it encompasses a broad variety of regulatory decisions. Chemicals are already regulated by many different statutes, including the Clean Air Act; the Federal Water Pollution Control Act; the Federal Insecticide, Fungicide, and Rodenticide Act as amended by the Federal Environmental Pesticide Control Act; the Occupational Safety and Health Act; the Marine Protection, Research and Sanctuaries Act; and the Safe Drinking Water Act. The report also considers regulatory problems that may arise under the proposed Toxic Substances Control Act. The problems we have considered may arise as "crises" resulting from public concern or the discovery of a potential hazard by an official agency; they may result from a methodical and systematic review of known potentially toxic substances; or they may result from legislative or judicial mandates for specific regulatory actions. The regulation being considered may involve direct control of the chemical product through such techniques as labeling or use restrictions; or it may involve control of the industrial sources of pollutants through restrictions on their discharge into the environment.

All the above distinctions and more that could be mentioned influence the nature of the decision-making process. Some of them will be used later in the report when specific aspects of decision making are discussed. But at this point the most important distinction to draw is between routine and nonroutine decisions. Nonroutine decisions are made by the top-level officials in an agency and may involve three types of actions: (1) setting broad policy direction for agency programs; (2) reviewing and approving standards and regulations which establish the criteria for specific lower-level decisions; or (3) resolving specific "cases" that are appealed to the agency head for final decision.

The vast majority of regulatory decisions made within EPA, the Occupational Safety and Health Administration, or other regulatory agencies are routine decisions such as reregistering a noncontroversial pesticide, granting an effluent discharge permit, or approving the occupational conditions within a specific plant. Most of the discussion and recommendations of this report, however, deal with nonroutine decisions because they have the greatest impact on society, and because they usually create the policies governing routine decisions. The internal bureaucratic organization, rules, and processes that distinguish nonroutine from routine decisions and that determine the level of the agency at which particular decisions will be made (and whether they will be made in the field or at headquarters) are themselves important aspects of the decision-making process, but we have not had the time or resources to investigate this aspect of regulatory decision making.

CHARACTERISTICS OF THE DECISION-MAKING PROCESS

Regulatory decision making within the federal government generally takes place under the pressure of tight deadlines and intense interest from concerned outside groups. These pressures sometimes hinder rational decision making, but their impact, particularly that of outside groups, is often helpful. In any case, such pressures are currently facts of life, and they are not likely to change in the future. Three other current characteristics of the decision-making process—the inadequacy (incompleteness or inaccuracy) of information, poor use of available information, and the necessity of making trade-offs—are also unlikely to change, but the way in which they are dealt with is subject to greater control by the regulatory agency.

All difficult decisions are characterized by inadequate information: if the decision maker knew all that he wanted to know, the decision would be less difficult, although painful trade-offs involving values might still have to be made. Problems of regulating chemicals in the environment are particularly beset with information characterized by a high degree of uncertainty. For some aspects of those problems there exists no information at all.

The quantity and quality of all types of information used for regulating chemicals must be greatly improved. The difficulty of ascertaining hazards to health and the environment posed by a chemical has been discussed at length in many reports and articles, but it is less frequently pointed out that the other kinds of information used in regulatory decisions may be just as uncertain or even more so. For example, a study of setting tolerance levels for hexachlorobenzene (HCB) showed that the decision made in this case was based in part on an estimate that between 17,000 and 22,000 head of cattle would have to be destroyed if the tolerance was set at 0.5 ppm. However, after the 0.5 ppm tolerance was officially established, farmers took various steps to modify the impact of the decision on their cattle, and only three cows had to be destroyed (Dominick 1974).

Even when adequate information exists, it may not actually be used in making the decision. Sometimes this may happen because the relevance of the information is not perceived. Sometimes it occurs because those responsible for the decision are unaware that the information is available. The statement made in 1972 by the OST–CEQ Ad Hoc Committee on Environmental Health Research that "communication between different groups of scientists, working in different disciplines and in different organizations, is imperfect and slow," is still true today (OST–CEQ 1972:30). We endorse a broadened version of that committee's recommendation:

Existing toxicological and other information systems related to the regulation of chemicals should be examined with a view to improving their coordination and use.

Scientific information concerning the hazard of a chemical which bears on a regulatory decision may come from a variety of scientific sources and disciplines to which the regulatory agency may not have easy access. An open meeting, well publicized in the technical press, can provide a forum for presentation of this information. Such meetings, attended by members of the academic community, the regulated industry, the regulatory agency, and the science press can provide a useful forum for presenting and evaluating current knowledge about the chemical in question. Such meetings can be particularly useful when new questions are asked about old compounds.

Information also may not be used because it is not "packaged" in a form which the decision maker can understand or apply. This has been a long-standing problem with respect to environmental monitoring data (NAS 1974). It is also a major problem because of the noncommensurability of many of the factors that should be considered in decision making, and this report will subsequently discuss ways in which the number of separate pieces of information can be reduced (see Chapters 6, 7, and 8).

A third characteristic of the decision-making process is the necessity of making trade-offs among conflicting values. The protection or establishment of some values always entails the sacrifice of others. Resources are always scarce, and their expenditure for one purpose means that other purposes will suffer.

The primary, although not the only, value that the regulation of chemicals attempts to satisfy is protection of human safety. But even if all other values were sacrificed to it, absolute safety would not be an achievable goal. Every chemical known to man would be fatal if one were exposed to excessive quantities of it. Operational definitions of safety can be established by prescribing various tests to be performed on a chemical, but there are always other tests that could be performed, and there is always the danger that the chemical may be misused. The decision maker is inevitably faced with the task of trading-off relative degrees of safety (or danger) against other values.

3 The Statutory Basis of Regulation

THE OVERALL STATUTORY FRAMEWORK

The statutes governing regulation of chemicals are a major determinant of both what will be regulated and how regulatory decisions will be made. A perfect process of decision making within a regulatory agency will not serve society's goals if the laws that are the basis of the process are defective. The law may focus attention on problems that are not of greatest importance, or it may mandate that the decision process consider only some relevant factors while ignoring others. Thus a consideration of the statutory basis of regulation is fundamental to an examination of decision making for regulating chemicals in the environment.

The current body of environmental laws does not represent a harmonious and purposeful whole. It was developed at different times by different committees of the Congress, and it reflects the vagaries of competing pressures and regulatory schemes. There are legislative actions that we believe could make federal regulation of chemicals more effective.

The passage of legislation similar to the proposed Toxic Substances Control Act would provide a systematic basis for developing and collecting needed toxicological and other information about industrial chemicals and would provide authority to regulate potentially hazardous chemicals that are not now subject to any statutory authority. It would thus fill a number of gaps in the existing regulatory structure.

The Working Paper on Market and Private Sector Decision Making, prepared for this study, proposes a scheme designed to identify and control toxic substances while retaining maximum freedom in the marketplace (see Appendix J). The paper suggests that specific criteria be developed requiring a manufacturer to evaluate the relative risks of a new substance. Producers would then have several options with respect to the distribution and labeling of the product. The regulatory agency would have the opportunity to review the appropriateness of the options chosen and to force a change to a more stringent risk category if warranted by the hazard potential of the chemical. The scheme has a unique element with respect to the assignment of liability. In the maximum risk category for marketed products, a tax is imposed and held in a trust fund against future claims for damages to persons, livestock, wildlife, or other components of the environment. The next most dangerous category, in which products are cleared by the government prior to marketing, has no tax on the product, but liabilities for damages are shared by the manufacturer and the government. In the other three less dangerous categories, the liabilities are assigned solely to the manufacturer.

We believe that this innovative proposal presents several serious problems. Uniform evaluation criteria are difficult to devise, because they run the risk of requiring unnecessary tests for some chemicals while failing to require essential tests for others. We are also wary about the effectiveness of liability as a means of exercising leverage in the regulatory process in a timely way. However, the proposal deserves serious consideration.

Two other gaps in the overall statutory framework require further exploration by EPA and Congress. One is the problem of indoor air pollution. The major part of a person's day is spent breathing the air inside buildings, and problems such as asbestos particles arising from heating and air conditioning ducts may have a significant impact on human health. Yet almost all of our regulatory effort (with the exception of occupational health efforts) has been devoted to cleaning the air outside buildings.

The other statutory gap is more complex and arises from the multiplicity of legislative authorities. It is an unfortunate fact that not all industrial or municipal wastes can be recycled or reduced to carbon dioxide or water. There are wastes that simply must be disposed of somewhere. The current statutory framework does not recognize this, except insofar as land disposal is less well regulated than disposal in other media and thus often becomes the most attractive way of getting rid of wastes. The air and water pollution statutes impose "blinders" on EPA that prevent the agency from looking at all the environmental effects of a pollution source at the same time. There is no established way within existing law that EPA or the states can rationally review a given waste disposal problem and determine what method of disposal would be least harmful to man and the environment.

EPA *should undertake a study to (1) identify neglected areas in hazard or pollution control, and (2) determine whether existing legal authorities are available and unused or whether new laws should be sought from Congress. Examples of such needs include the problem of choosing optimal waste disposal methods considering the total environment, and the question of indoor air pollution.*

INCONSISTENCIES IN THE STATUTES

Given the piecemeal way in which environmental legislation has developed, it should be no surprise that not only are related acts inconsistent, but inconsistencies and ambiguities exist between different sections of the same act. That a particular chemical may sometimes be handled differently under different statutes is not necessarily illogical, because different circumstances may result in quite different costs and benefits. A chemical such as arsenic, for example, may be treated one way when used as a pesticide but another way when emitted into the air or water as an industrial by-product. However, there are several aspects of the overall statutory framework which are of critical importance and for which a greater degree of consistency is desirable.

BURDEN OF PROOF

Because the scientific evidence regarding health and environmental effects is so difficult to obtain with precision, and because the costs of data collection can be so high, the party carrying the legal burden of proof is at a considerable disadvantage. The issue of who should have to prove the safety or hazard of a chemical is complex, for the burden of proof can shift during different stages of the administrative process and can require different degrees of proof.

The clearest statutory burden is in the pesticide area, where agricultural chemical companies must demonstrate affirmatively that their product is "safe." However, in most environmental legislation specification of the party that must prove the "safety" or hazard of a chemical is left vague and has to be established on a case-by-case basis by the courts.[1]

The organization that proposes to build a project or market a new chemical is the one that plans to benefit from it and, usually, the one with the resources to analyze its unfavorable as well as its favorable consequences. It is also in the best position to insure that societal costs are reflected in the price of the product. The recent trend is away from acquiescence by society in the acts of individuals and towards placement of the burden of proof on those who propose to carry out a project or market a product with probable environmental consequences. We believe that this is an appropriate and healthy trend.

As a general principle the burden of proof that society will obtain a benefit from a new *use of a chemical should rest with those proposing such use. It may be desirable to make statutory changes to reflect this principle.*

The application of this general principle to manufacturers intending to market a new chemical is clear. It is less clear in cases in which the government proposes to regulate a chemical already on the market. However, we believe that the same principle should apply to the continued use or marketing of a chemical; and thus, once the government has made a reasonable case that the use of the challenged chemical creates an excessive hazard to human health or the environment, the burden of producing evidence should shift to the proponent of use.[2]

Once the government has made a reasonable case that the challenged use of an existing *chemical creates an excessive hazard to human health or to the environment, the burden of producing evidence should shift to the proponent of use, who must then make an appropriate showing that the continued use is desirable.*

DISCRETION LEFT TO THE AGENCY

Currently there exists wide variation in the discretion delegated to administrative agencies by legislation regulating chemicals. The spectrum reaches from the Delaney Amendment of the Federal Food, Drug and Cosmetics Act [FDC Act, Sec. 409(c)(3)(A).], which flatly prohibits the use of food additives shown to be carcinogenic in appropriate animal experiments, to the almost total discretion granted EPA under the Clean

[1]Safety and hazard are inherently subjective categories, but they can be given specific meaning through statutory provisions and administrative regulations.

[2]Congress must direct the agencies and the courts as to what constitutes a reasonable case and an "appropriate showing." It may consist of "substantial evidence," "preponderance of the evidence," or substantiation "beyond a reasonable doubt."

Air Act to regulate fuel additives (Clean Air Act, Sec. 211.) and to set the time for achievement of secondary air quality standards (Clean Air Act, Sec. 109.).

A number of conflicting values enter into any consideration of how much discretion should be left to the agency. These include: (1) Predictability—specific legislative standards are predictable, at least to their authors, while standards to be derived from delegated authority are not. However, predictability can often be eroded in implementation, as, for example, in the choice of testing methods to be used in enforcing the automobile emission standards in the Clean Air Act. (2) Promptness—administrative agencies may be slow to act without a statutory or other external stimulus. (3) Complexity and volume—certain types of regulatory decisions are either sufficiently complex scientifically or so frequently recurrent but factually distinct (pesticides registration and issuance of effluent discharge permits are examples) as to require administrative, as opposed to legislative, regulation. (4) Flexibility—the more likely a situation is to change, the more discretion should be left to the agency, because it is often difficult to obtain rapid changes in a statute. (5) Legitimacy—public acceptance of any standard is essential, and the public may regard legislative regulation as more legitimate than administrative regulation. The more a regulatory decision is viewed as setting important national policy, the more it is considered appropriate for Congress to make the decision. (6) Accountability—the reason that congressional decisions are regarded as more legitimate is that the Congress, in contrast to the bureaucracy, is more politically accountable to the people. On the other hand, decisions made by an agency, unlike most congressional decisions (except those that raise constitutional issues), are ultimately subject to judicial review. Thus, in a practical sense, the public may have more control over administrative decisions than over congressional decisions.

As a general principle, the more far-reaching the regulation, the more appropriate it becomes for legislative determination, at least regarding policy direction. The more uncertain the scientific basis for regulation, and the greater the need for flexibility and adaptability, the more administrative regulation is appropriate. However, given the diverse values involved, no uniform rule can determine the degree of agency discretion to be allowed.

WEIGHT GIVEN TO HEALTH VERSUS OTHER FACTORS

One aspect of the discretion problem—the weight given to economic factors in regulatory statutes designed to protect health—is of sufficient importance to warrant separate consideration. Several important statutory provisions, such as those in the 1972 Amendments to the Federal Water Pollution Control Act dealing with toxic effluent standards [Sec. 307(a)] and the Clean Air Act provisions for establishing primary air quality standards, have been interpreted to require that economic factors be ignored in decision making.

We believe generally that such provisions in the law are unwise. All factors in a decision, including economic factors, should be considered in an orderly manner. The weight given to the various factors may, however, be very different in different cases. In some types of situations it may be desirable to give predominant weight to the health or environmental effects of a chemical, whereas in other cases considerations of health, the environment, economic impact, and other factors may all be given equal consideration. Legislation that narrows the items to be considered by regulatory agencies in reaching a decision weakens rather than strengthens the decision process and frustrates those who seek to make deliberate decisions concerning the environment. Also, the public, as well as government officials, may accept the concept of a "socially acceptable risk," particularly for chemicals that are in widespread use and give large benefits to society. Finally, practicality dictates that economic factors be considered in the decision-making process, because even if they are not considered explicitly they will almost inevitably have an influence on the final decision. For example, the establishment of widely varying margins of safety in the primary air quality standards reflected, among other factors, the economic costs of achieving the standards.

Statutory provisions should not preclude consideration of any relevant factors in the decision-making process. Provisions that prevent such consideration should be evaluated for possible amendment.

Having stated this general principle, we would add that additional guidance from Congress to the agencies on the weight to be accorded various factors in decisions would be desirable. Such explicit guidance is exceedingly difficult to give, but it would improve the consistency, predictability, and accountability of agency decisions, provided that it did not exclude important factors in the decision.

Whenever possible, and without precluding administrative consideration of relevant factors, Congress should provide increased and consistent statutory guidance as to the relative importance that should be given to health, environmental, and economic factors in regulating chemicals.

JUDICIAL REVIEW

Most chemical regulatory statutes are silent on what the appropriate standard of judicial review should be. Yet this determination can be crucial in deciding an issue. To set specific standards, Congress must make a value judgment as to the proper role of administrative discretion. In quasi-legislative administrative proceedings, such as rule making, a "preponderance of the evidence" test may be required; in adjudicatory decisions, which have built in the procedural safeguards of the legal system, a substantial evidence or "arbitrary and capricious" standard is often applied. In any case, to increase the predictability and consistency of judicial review, Congress should articulate the standard of judicial review it intends.

The trend toward an ever greater role of the courts in chemical regulation and toward greater reliance on the appeals process is unfortunate, because it greatly lengthens the time until a decision becomes final, and because the courts inevitably possess less knowledge about the technical aspects of chemical regulation than executive agencies. If the initial decision process for which an administrative agency is responsible is based on adequate and reliable information and on broad participation, it will be more likely that the ultimate issue will be decided before it reaches the courts. The recommendations given throughout this report should help to achieve this end. In addition, there may be steps that could be taken that are specifically aimed at reducing the tendency to resort to the courts, such as holding public hearings at an early stage in the decision-making process.

The early and open exchange of information and opinions on a proposed decision should be encouraged in order to reduce the current dependence on subsequent judicial challenge. The EPA *Administrator should hold public hearings at the earliest feasible stages of the decision process. He also should facilitate prehearing exchange of information among parties (for example, through depositions, interrogatories, and other discovery procedures).*

Once a case reaches the courts, its outcome may depend heavily on the knowledge and values of the judge. But neither the judge nor the lawyers presenting the case to the judge are likely to be technically trained in the disciplines (such as economics or health and environmental sciences) relevant to deciding regulatory cases involving chemicals. The primary focus of the judge is on questions of legal procedure, in which he is an expert, but it is often difficult to make a clear separation between legal procedure and substantive questions about the effects of a chemical on health, the environment, or the economy. There is a need to examine the problems of making technical decisions in a judicial context.

EPA *should ask the American Bar Association to cooperate in a study of the way health and environmental information and economic analyses should be introduced and considered in judicial proceedings.*

ORGANIZATIONAL STRUCTURE

Although it seems clear that the organizational form of the regulatory agency will affect the way it carries out its regulatory responsibilities, there is little agreement about the kinds of effect that particular forms of organization have. Some members of the committee thought that a multimember independent commission to regulate chemicals would be more responsive to all interest groups and would also be able to make better use of technical information. Other committee members thought that such a structure would dilute responsibility and responsiveness and favored the current arrangement of a single administrator serving at the pleasure of the President.

Within the Committee there were divergent opinions on the optimum organizational structure for making chemical regulatory decisions. It is recommended that Congress give consideration to alternative decision-making structures. In particular, it should consider whether responsibility should be vested in a board or commission with fixed terms in office or in a single administrator.

4 Advocacy and the Regulatory Process

OPENNESS IN DECISION MAKING

EPA and, to a lesser degree, other regulatory agencies have made in recent years a concerted and praiseworthy effort to make the decision-making process more open and accessible, but it can be made still more equitable. Equity requires that the costs and benefits of the decision be fairly, although not necessarily equally, distributed. It is often useful to shift the focus from fairness in the distribution of outcomes to fairness in the decision process itself. By doing so, more agreement is likely to be reached about what is fair; and the best way to promote fairness in the outcome is also to establish fairness in the allocation and use of resources employed to reach the decision. Thus we propose as a useful principle of equity that the decision process is equitable when the consideration given to the interests of potentially affected individuals is proportional to the anticipated effects of the decision on those individuals. In other words, the decision process should include the groups in society who will feel the impact of the decision, and it should reflect the degree of impact of the decision on those groups. Because any major regulatory decision about a chemical is likely to affect many diverse groups in society, the decision-making process should be as open and accessible as possible.

Practical considerations also call for an open decision-making process As will be described in more detail below, the information required to make decisions often must come from nongovernmental groups. The more the information is exposed to critical review by a diversity of interests, the more likely it is to be reliable. Neutral disinterested research has a vital role to play in the regulation of chemicals; but unless such research can be scrutinized by all the various parties involved in a decision, there is no way of being sure that it is neutral and disinterested.

Once a decision is made, its implementation depends heavily on public acceptance. The day when a decision was accepted as legitimate simply because it was made by a government agency has passed, and the process by which a decision is arrived at has become perhaps the most important determinant of its acceptance by society. The more open the decision-making process, the more feasible and effective will be the implementation of the decision.

THE SCIENTIFIC COMMUNITY

A major concern about an open decision-making process is that it will excessively politicize the process and thus minimize the role of scientific expertise. We believe, however, that openness and scientific input tend to be complementary rather than contradictory. Many of those with scientific knowledge relevant to a decision are not part of the governmental apparatus, and their knowledge is more likely to influence the decision if the process is open to all outsiders. Also, we believe that the reliability and acceptability of scientific evidence is likely to gain from critical scrutiny.

The scientific and technical quality and the policy relevance of each major research project supported by a regulatory agency—whether to be performed in-house or by contract—should be reviewed in advance, to assure that it is appropriate to achievement of the agency's mandates and goals; and again at the conclusion, to evaluate how effective and timely the results have proved when considered against the initial purposes and schedule. The findings of this review group should be published at least annually.

It is of the utmost importance that regulatory decisions reflect the best scientific knowledge available. Several steps can be taken to help insure that this is the case.

All agencies that regulate chemicals should establish procedures for external scientific review of the technical data base presented to the decision maker. The results of the review should be available directly to the decision maker and to the public as early as feasible. Specifically, the major research and development programs undertaken by EPA *should be reviewed routinely by panels of qualified experts.*

EPA has recently established a Science Advisory Board (SAB).[3] The SAB should have a permanent list of scientific consultants so that the scientific basis of all important decisions can be reviewed rapidly by the foremost experts in a relevant area. If the SAB thinks that additional data, controlled by a nongovernmental source, are essential to a valid analysis of a given problem, the Board should have a formal procedure for requesting this information from the source. If the data are still withheld, the Board should be able to request that the decision maker exercise his authority to subpoena the information.[4]

EPA*'s Science Advisory Board should be permitted to formally request the Administrator to subpoena information needed for the Board's consideration and to allow the Board to investigate new areas. The request and the Administrator's response should be a matter of public record. The existence and responsibilities of the Science Advisory Board should be mandated by legislation.*

We are aware that outside scientific boards can become counterproductive, especially if there is not a clear definition of where review of technical analysis ends and application of sociopolitical judgments leading to a decision begins. However, if such a board can perform its task well, avoiding attempts to make decisions or preempt the administrator's role in decision making while maintaining independence in exercising its technical judgment, it can serve a vital role in the decision-making process.

Aside from the increased use of scientific review and advisory boards, there are other steps that can be taken to provide adequate and reliable scientific information for the regulatory process.

To improve the quality of scientific evidence, the right of counsel to cross-examine expert witnesses should be guaranteed.

The power to subpoena expert witnesses, now available to EPA *under the Federal Environmental Pesticide Control Act, should be extended to proceedings under other statutes to develop the fullest and fairest public record.*

[3]Technically the new EPA Science Advisory Board consists of all members of a number of existing EPA scientific advisory committees (the Ecology Advisory Committee, the Hazardous Materials Advisory Committee, and so on). The real functions of the SAB are likely to rest with its Executive Committee which will consist of five to twenty members. Thus when we refer in the text to the SAB, we mean more precisely the Executive Committee of the SAB .

[4]See the Working Paper on Regulatory Options, Appendix I, for further details.

INDUSTRY, LABOR, AND ENVIRONMENTAL GROUPS

A wide diversity of resources exists among the many groups affected by a decision. Efforts are necessary to improve the balance between the anticipated effects of a decision and the resources available to influence the decision. In the regulatory process information is power, and an imbalance in the resources to obtain information can lead to both inequitable and inefficient regulatory decisions.

The producer sector of the economy generally is better equipped to protect its interests than is the consumer sector, and our recommendation on burden of proof reflects this fact. But within the producer sector, small firms are often severely handicapped in trying to influence decisions that affect them.

The Department of Commerce should develop an educational resource to help small businesses acquire the information on chemical regulatory matters that is at present routinely available to large corporations and major trade associations.

Industrial workers are generally recognized as bearing a disproportionately high risk of exposure to toxic materials, because pollution inside a factory is frequently more severe than outside. Those workers who live near the factory frequently suffer from "double jeopardy," because both their work and nonwork environments are poorer than the national average. Labor is also the segment of the community most vulnerable to the economic consequences of controlling toxic materials (for example, when controls might mean shutting down a factory). Most workers in the U.S. are not members of unions, and several million of those who are organized belong to small independent unions that do not have significant professional resources. Therefore, efforts should be made by regulatory agencies and by the large unions to insure that the interests of all workers affected by a decision are represented.

The contribution of environmental and consumer groups to the improvement of decisions has been recognized both within and outside the government. In a recent speech, the EPA Administrator told one of the largest conservation groups, "Your organization . . . has done much to alert America to the dangers of environmental abuses. EPA welcomes your support, and we will need your help in the years ahead" (Train 1975:10). The recommendations made below to increase the openness of decision making would support and augment the role played by such groups.

THE NEWS MEDIA

The news media have the capacity to inflame public opinion or to calm it, to inform the public or mislead it. The government is in a position to help guide those capabilities through a policy of candor and frankness, taking the press into its confidence at various stages in the decision-making process.

Reporters generally pride themselves on their independence and insist on the right to cover a story as they see fit. There is frequent opportunity for the government to communicate freely and frequently to reporters—both in formal briefings and in one-to-one coverage—what it is doing to protect the public with respect to chemicals in the environment.

Regulation of toxic substances is, by its very nature, a subject that will evoke sympathetic reactions from the press. It is a story with a built-in advantage for the government which is, after all, engaged in the business of protecting the same public the press sees itself as representing.

The press as well as other interested parties should be told when a standard-setting process begins and when the proposed standard is ready for publication; further discussion during this process should occur as often as warranted.

STATE AND LOCAL GOVERNMENTS

The general trend of environmental legislation and regulatory activities in the toxic substances area is clearly away from the direct involvement of state and local units of government and toward a preeminent federal role. Although it did not explore the desirability of this overall trend, the Committee believes that state and local governments can be used more effectively as sources of information for regulatory decision making.

State and local health and environmental agencies often possess relevant data that are important to federal decisions and that are not available from any other source. For example, state environmental agencies as well as state departments of commerce and industry will often know in great detail which industrial facilities in a state might be affected by the potential regulation of a chemical. Also, state agencies may serve as a useful conduit for tapping the knowledge and information held by nongovernmental experts within the state, particularly those in state universities.

Federal agencies regulating chemicals should establish a formal and regular method for obtaining information from state agencies.

OTHER RECOMMENDATIONS FOR OPENNESS

There are a number of further steps that could be taken which would make the decision-making process more open and equitable.

One set of recommendations is designed to insure that the information and considerations that are relevant to a decision are publicly available to the maximum extent. This is a prerequisite for effective public participation.

Any information available to an agency on the hazards of a chemical that is regulated by that agency should not be considered proprietary and should be available for public inspection in a timely fashion during and after the decision-making process.

Specific nonproprietary information on hazards includes not only data on the intrinsic toxicological properties of a given substance, but also data on patterns and quantities of use when such information is essential for the evaluation of the overall hazard of a given chemical. Certain use data (e.g., to whom a chemical is sold) will remain proprietary, unless that data can be shown to be essential to evaluation of the hazard.

The essential elements of decision making should be part of the public record. The agency should publish a "white paper" for each important regulatory action undertaken. The paper should include the key details of the economic, legal, scientific, and other considerations taken into account in reaching the decision. It should be issued when the agency decides to take some action but sufficiently in advance of a final decision to permit considered response. An important decision to take no regulatory action, or to defer such action, should also be accompanied by a "white paper."

The white paper should outline the available options and discuss the reasoning that the agency followed in choosing the one it did. It should also demonstrate a recognition of supporting or conflicting views offered by the public and by other government agencies. We foresee this as being similar in purpose and concept to the environmental impact statements prepared in accordance with the National Environmental Policy Act of 1969, although the white papers would be broader in scope. The inflation impact statements now being required by the Executive Branch also suggest the need for a procedure of the type we have recommended. The preparation of white papers will require significant budgetary and other resources, but we believe that the papers are of sufficient importance to warrant such expenditures.

All ex parte *communications, including those from Congress, members of the Executive Branch, private corporations, and citizen groups, on any adjudicatory decision pending before the agency should be made public with sufficient time for comment before a decision is made.*

Communications from Congress or the White House presumably may have a significant influence on a decision, and thus this recommendation is consistent with the principle of making public the elements that are important in reaching a final decision.

A second set of recommendations is designed to deal with what the Committee perceives as inequities between the allocation of resources to influence the decision and the degree to which parties are affected by the outcome of the decision.

The purposes of creating a more open and equitable regulatory process are to improve the quality of information used in decision making, to lessen the current burden upon the agency of representing the otherwise unrepresented interests (including future generations), and to prevent EPA from being "captured" by the interests it is supposed to regulate.

At appropriate points in the decision-making process the agency should actively seek to identify the affected parties and solicit suggestions and comments from them. Ways should be explored to better represent the interests of future generations.

Publication of a proposed action in the *Federal Register* is not an adequate method for soliciting the comments of those who may be affected by a decision. Informed participation of the news media in eliciting public cooperation and comment is highly desirable. Special lists of nonfederal organizations with particular interests should be used to invite their response.

EPA *and other agencies should initiate programs aimed at training and encouraging citizens to participate in the decision-making process.*

Although the right of government to exert political pressures by manipulating the anxieties of various groups is correctly questioned, there should be no question of the duty of government to go beyond the public notice and the press release to identify and assist social structures in coping with technological impacts. It is necessary to train field workers for this task. They must not themselves "skew" the information, arbitrarily favor one group over another, or become organizers of citizen action. Rather they must find leaders of every kind who have an interest in the issues, provide opportunities in training for the citizen role, and then maintain a flow of information on the facts, issues, and opportunities for participation. The now extinct community support program of the former National Air Pollution Control Administration might serve as a model for such a program.

It might be argued that more openness and participation will greatly delay reaching a final decision. There is agreement that many significant decisions have taken far too long to make, but this has more often been due to parties with a specific interest in delaying a decision than to public participation per se. We believe that steps, such as specifying a deadline (which the administrator could extend only for good cause) or establishing a penalty as a disincentive for any party to delay unreasonably, can be taken to reduce delay in the decision-making process without in any way sacrificing openness or accessibility.

Congress should review the adversary procedures that have led to unduly protracted hearings (such as pesticide cancellation) and determine how best such marathon decision making can be shortened.

COORDINATION AMONG FEDERAL AGENCIES

Despite the theoretically hierarchical structure of the Executive Branch, from the standpoint of any one agency the other federal agencies are external and sometimes competing forces. Constructive interagency competition is useful in obtaining information and opinions about various regulatory options. But there is a need for greater cooperation among agencies with respect to both information and policies. EPA's relationship with other federal agencies concerned with the agency's regulatory actions should be strengthened substantially.

Other federal agencies should have the opportunity to participate in internal EPA *deliberations early in the consideration of regulatory options. The same opportunity should be given to* EPA *by other agencies that regulate chemicals.*

When an important new problem is identified, a great variety of information may be required in a short time. Often, it can be obtained more expeditiously and more completely by a coordinated multiagency program, with different parts of the problem delegated to the agency in the best position to complete the task in the time required. Active interagency cooperation is crucial to many agency research programs.

Agency cooperation with respect to regulatory policies or actions is necessary either to obtain the optimum degree of consistency among different agencies dealing with similar problems or to settle conflicts when more than one agency has jurisdiction over the same problem. For example, the jurisdictional overlap between EPA and the Food and Drug Administration on pesticide residues has been partially resolved by memoranda of understanding between the two agencies, but the relationship between EPA and the Consumer Products Safety Commission on many issues involving chemical regulation remains clouded.

One significant but neglected area requiring interagency coordination involves the allocation among different regulatory authorities of the total allowable exposure to any particular chemical or other substance.

Exposure to many chemicals comes simultaneously through food, air, occupational exposure, and other routes. At the present time the regulations of each agency are based largely on the assumption that the exposures regulated by that agency are the only exposures. Thus, it is quite conceivable that an individual could receive several times the safe exposure even though all regulatory standards were enforced adequately. In addition, no agency oversees the possible synergistic effects of regulated chemicals.

Exposure to many chemicals comes simultaneously from two or more media. An interagency committee should be established to ensure that each agency's standards reflect appreciation of such multiple exposures.

The "white paper" discussed above, or the environmental impact statement if one is filed, could be used by the committee to check on the degree to which an agency considered multiple exposures in setting a standard. This is another advantage of making the basis for a regulatory decision public and explicit.

5 The Decision-Making Process

WHO MAKES THE DECISION?

A primary consideration in examining the decision-making process within an agency is who in the agency should make the decision. An essential aspect of this question is where in the agency's hierarchy the decision should be made. Another important aspect is the extent to which analysis done at the lower levels within the agency should determine the outcome of decisions made at the top level.

As noted in Chapter 2, the vast majority of regulatory decisions are "routine" and involve the application of policies established by the top-level decision makers to a particular situation. However, the principal focus in this report is on regulatory decisions that require the direct policy involvement of the agency head or other high-level personnel. There is almost always some degree of discretion and judgment involved even in routine decisions, and thus whether top-level decision makers become involved depends largely on the degree to which the decision is determined by already established general policies and on the magnitude of the socioeconomic consequences of the decision.

A more subtle but no less important question relates to the analytical staff work on which the top-level decision maker relies for making nonroutine decisions. It frequently happens that a decision nominally made by the agency head is in fact made by lower-level technicians or support personnel because the analysis presented to the agency head leaves him only one feasible choice. Occasionally there may really be only one choice that makes sense, but the more usual case is that the technicians have knowingly or unconsciously mixed technical findings with sociopolitical judgments in presenting options to the agency head.

Because the bureaucratic staff serves free of any political accountability, we believe that the decision-making system should minimize the extent to which the technical experts are allowed to superimpose sociopolitical value judgments on the technical data base. As will be made clear below, judgment is an inescapable element in making any nonroutine decision.

The decision process should require the agency's technical staff to present a full set of options with a corresponding range of cost–benefit–hazard data and explicit statements on the confidence limits of each analysis.

NEED TO SET PRIORITIES

To a great extent the current agenda of EPA and other regulatory agencies is determined by reaction to outside events. Stories in the morning newspaper probably have more impact on what decisions come before the agency head than most internal agency processes of problem identification or priority- setting. The agency must always be sensitive to current public concerns, but we believe that a more forward-looking, planned set of priorities would be desirable.

EPA is devoting significant resources to planning efforts. We have not explored the quality of this work or the extent to which it is incorporated in the regular internal workings of the agency. However, to supplement these efforts, greater use of outside experts would be helpful.

EPA *should make greater use of committees of experts who represent the spectrum of potential viewpoints, particularly in helping to anticipate future chemical control problems.*

Another step that would help in setting rational priorities is to use a matrix or generic approach so that when one problem comes to the Agency's attention, other closely related problems are also reviewed for possible action. For example, when a hazard from a particular pesticide is determined, EPA should attempt, to the extent its resources permit, to examine the other members of that pesticide class at the same time. To investigate the hazards of PCBs without examining polybrominated biphenyls is both inefficient and needlessly risky.

EPA *should adopt, whenever scientifically possible, a generic approach, as opposed to an ad hoc procedure, for the regulation of chemicals.*

In addition to these proposals, the hazard rating scheme discussed later in this report would provide a valuable tool for establishing priorities. It makes no sense to go to enormous effort and expense to reduce the amount of a chemical in the environment so as to save a few lives if a similar effort could be directed at a different chemical and result in saving thousands of lives. The value attached to saving a human life and to other types of hazard reduction should be roughly comparable for different decisions, and a system of hazard ratings would be a step in this direction.

LEVELS OF SOPHISTICATION AND SEQUENCING IN DECISION MAKING

The sequence of steps leading to a regulatory decision is of vital importance in determining the quality of the decision. The steps to be taken and the effort expended by the agency on each step will be determined by the gravity of the problem and the resources available to the agency.

The initial determination must be whether the chemical in question poses a hazard to health or the environment. However, this question can rarely be answered unambiguously. Almost always the key questions relate to the likelihood and magnitude of the suspected hazard. The next logical question is whether the production of the chemical produces benefits. Because benefits to individuals as well as to society must be considered, the answer to this question is almost invariably yes. However, if a chemical is not hazardous or not beneficial, then clearly the regulatory decision is simple.

Because the conditions for simple decisions are rarely met, more sophisticated sequences of decision-making steps must be used. An outline of such a sequence is shown in Figure 1. Some of the key concepts in this sequence will be explained later in this chapter.[5]

For any sequence of decision making it is possible to set forth characteristics and objectives of the process which, if followed, will maximize the usefulness of the process for the decision maker. These characteristics and objectives are as follows. (1) The decision process should begin with a model, picture, or flow diagram of the total system of production, distribution, use, and fate of the chemical. (2) The decision-making framework should make it possible to identify and present information on the range of alternatives the decision maker has. (3) The framework should include the major identifiable effects and consequences of alternative actions. (Chapters 7 and 8 of this report are devoted largely to an examination of how this information should be collected and aggregated.) (4) The framework should facilitate the

[5]The full sequence is explained in the Working Paper on Regulatory Options (Appendix I). Also see NAS (1975b:24).

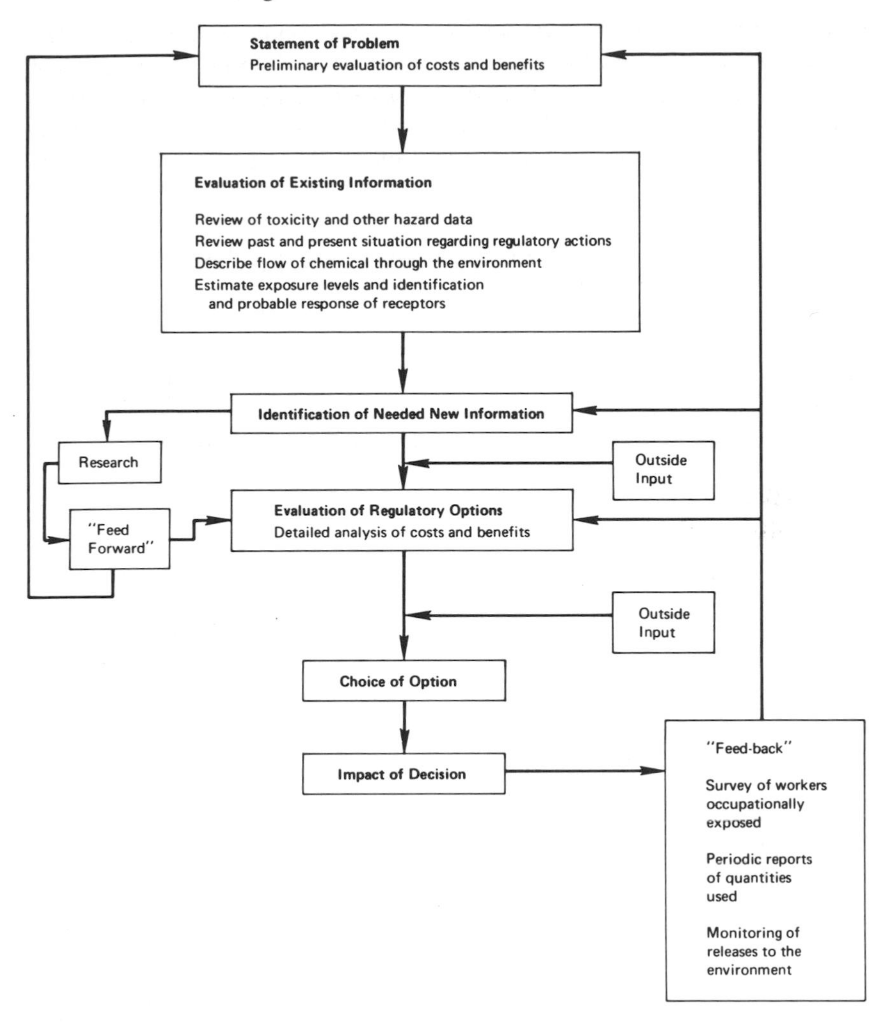

FIGURE 1 Sequence of steps for nonroutine decision making.

comparison of major effects resulting from the alternative actions and should serve as a convenient basis of discussion and review. (5) The framework should be designed so that it can be tailored to the kind of decision to be made and to the needs of the decision maker. The form in which data are presented is itself in part a policy decision. It should not be treated as a purely technical matter and should be a conscious choice of the decision maker. (6) The framework should make it convenient to determine the value of obtaining further information and, specifically, which information should be obtained. (7) For most analysis there will be a great deal of uncertainty about some of the data and thus about the effects of different regulatory options. The framework should indicate the range of uncertainty and levels of ignorance about key pieces of information. (8) The framework and decision process should permit sensitivity analysis for variations in both input data and different assumptions about relationships. Subjective judgments and assumptions made by the analysts that significantly affect the estimates of effects should be made apparent to the decision maker. (9) To the greatest extent possible, all effects should be quantified. In addition, there should be as few measurement categories or units as possible to facilitate aggregation of effects when necessary and to simplify the trade-off considerations of the decision maker.

FEEDBACK AND "FEED-FORWARD"

Feedback information is that which flows back to the initial decision maker following the execution of a decision and the occurrence of impact. "Feed-forward" information is data that result from an effort, as part of an initial regulatory decision, to construct a more complete and validated data base upon which to evaluate future decisions in the same area. Both kinds of information are vital for the optimal performance of any decision-making system, yet neither kind is adequately provided for at the present time. Formal procedures to ensure that they are included in the decision-making process should be developed and institutionalized in EPA and other agencies.

In the course of making a decision, gaps or weaknesses in the information and data base should be clearly identified and steps taken to provide for feedback information as the basis for subsequent modification of the decision. In some cases the agency should commit itself specifically to perform a formal periodic review of the information, particularly if the decision has been to allow the use of a "suspect" product or the continued emission of a pollutant. In many cases the responsibility for collecting feedback information should be placed on those who want to initiate the proposed action. When another government agency is a proponent (for example, when the U.S. Forest Service wants to use a pesticide on the national forests), that agency should be responsible for collecting feedback information and should also bear the cost of feed-forward information gathering.

As regulatory options are being considered for a given problem, there should be a procedure to ensure that a specific assessment is made of future needs for information associated with each option considered.

There have been few examples within EPA of adequate provision for feed-forward information, and the efforts that have been undertaken have been organized essentially on an ad hoc, case-by-case basis. One example where adequate provision was made was the case of the November 1973 decision to allow catalytic converters on automobiles to control exhaust emissions. As part of this decision, the EPA Administrator initiated a high-priority research program to determine whether the concerns voiced by critics of the catalysts were valid. The research that was done has been subjected to considerable criticism (NAS 1975c), but it was of vital importance in the March 1975 decision to extend the deadline for meeting more stringent auto emission standards. Our point here relates not to the quality of the research nor to the validity of extending the deadline but rather to the need for systematic planning, as part of the ongoing regulatory process, of research required for making future regulatory choices.

Clearly there are political and other penalties imposed on an agency or decision maker who reverses or modifies a previously made decision. But the penalties to the public, and ultimately to the decision maker, of not reconsidering past decisions are even more severe. The regulatory agency must assume a firm responsibility to support research where appropriate and to bring new science to bear on old decisions. There must be conscious and continuous review of the basic assumptions and methodology of past decisions so that future mistakes can be avoided.

For optimal regulatory decision making, a procedure to conduct retrospective analyses of the impact of given decisions should be adopted. This should emphasize evaluation of the accuracy of the predictive models for health, economics, and environmental sciences in the original decision-making process. EPA *should also develop a formal system that reviews and identifies information necessary for optimal future decision making and implements the appropriate research programs to generate this information in a timely fashion.*

The pluralist nature of research support in the federal government can provide another check on the decision maker who ignores new data or who is reluctant to support new research which questions old decisions. Other federal agencies that perform research of a more basic nature (HEW, for example) or for a different purpose (the Energy Research and Development Administration, for example) should be able to provide the checks and balances necessary to prevent the decision maker from inhibiting or slanting the direction of new research.

6 The Uses and Limits of Benefit–Cost Analysis

PERSPECTIVE

There is no objective scientific way of making decisions, nor is it likely that there ever will be. However, use of the techniques developed by decision theory and benefit–cost analysis can provide the decision maker with a useful framework and language for describing and discussing trade-offs, noncommensurability, and uncertainty. It can help to clarify the existence of alternatives, decision points, gaps in information, and value judgments concerning trade-offs. Furthermore, it should facilitate communication beween the decision maker and his staff of analysts, and between the decision maker and the public.

Traditional benefit–cost analysis is an application of economics that has been used to analyze various kinds of decisions, particularly large investments in public goods.[6] Decision theory, although in some ways closely related to benefit–cost analysis, has developed as a separate discipline. It differs from benefit–cost analysis in its emphasis on tracing through the consequences of each of a series of sequential decisions, in its substitution of the values of the decision maker for market prices, and in its use of probability theory to deal with the problem of uncertainty.[7]

[6]For discussions of benefit–cost analysis see, for example, Eckstein (1958), Dorfman (1963), Hitch and McKean (1965).

[7]For discussions of decision theory see, for example, North (1968), Pratt et al. (1965), Tribus (1972).

For reasons explained below, we have found traditional benefit–cost analysis to be not very useful in making decisions about regulating chemicals. The analytical tools of decision theory do provide a useful way of assembling or displaying information, but we do not believe they can realistically be used to make the final decision. If for no other reason, decision theory is limited because top government executives will simply not take the time to assign quantitative values to the varied elements that enter into any particular decision.[8]

In this report, we have used the term "benefit–cost analysis" in a loose sense to refer to the body of knowledge and techniques belonging to both traditional benefit–cost analysis and decision theory. When strict benefit–cost analysis (as used, for example, by the Army Corps of Engineers) is meant, we have referred to "traditional" benefit–cost analysis.

Benefit–cost analysis, at least as we use the term, is not a rule or formula which would make the decision or predetermine the choice for the decision maker. Rather, it refers to the systematic analysis and evaluation of alternative courses of action drawing upon the analytical tools and insights provided by economics and decision theory. It is a framework and a set of procedures to help organize the available information, display trade-offs, and point out uncertainties. In this way benefit–cost analysis can be a valuable aid; but it does not dictate choices, nor does it replace the ultimate authority and responsibility of the decision maker.

We shall first describe the limitations inherent in any type of systematic analysis of chemical regulatory decisions. These limitations preclude the substitution of analysis for conscious, responsible choice by decision makers. We will then discuss the ways in which analysis can assist in making responsible decisions.

LIMITS OF BENEFIT–COST ANALYSIS AS A DECISION TOOL

VALUES

The most important and pervasive limitation on benefit–cost analysis is the role of values. Many of the factors that are likely to be most significant in a decision concerning toxic chemicals cannot be measured

[8]They may also be unwilling to commit themselves to the logical conclusions that follow from the assignment of utility values.

in common terms (such as dollars) that are agreeable to all concerned parties. Different individuals place different values on things such as human life, aesthetics, or national security. Thus, an analysis that assigns a quantitative value to one or more of these factors is necessarily subjective and, to some degree, arbitrary.

Consider a hypothetical and highly simplified case of a pesticide in current use. Assume that the following information is known with certainty. At current levels of use, the pesticide increases the net agricultural productivity of the land on which it is applied by $500,000 per year; but it causes 10 deaths per year. If usage were reduced 50 percent by regulation, productivity would fall by $200,000 per year and five deaths per year would be avoided.

This information can be displayed as follows:

Level of Use	Benefits of Use (Productivity, $/yr)	Costs of Use (Deaths/yr)
current level	500,000	10
50 percent control	300,000	5
ban or zero use	0	0

The problem is clearly one of trade-offs between productivity and human life. But no simple decision rule can be applied here, because there is no objective basis for making deaths and productivity in dollars commensurable. No objectively determined weights or values have been provided for converting these noncommensurables into a common unit of measure.[9] The outcome of the decision will depend on the value that the decision maker places on human life.

Although there is no objective way of deciding the value of a human life, different decisions inevitably imply different values. In the above example, if the decision maker chooses the 50 percent control option, he is implicitly saying that he is willing to give up $200,000 in productivity in order to save five lives, i.e., lives saved are worth at least $40,000 apiece in the decision maker's scheme of things. But because the decision maker chose not to go to a total ban, this reveals that he was not willing to give up an additional $300,000 to save five more lives. Thus his implicit, and perhaps subconscious, valuation of a human life is less than

[9]A number of attempts have been made to establish objective ways of placing a dollar value on human life. See, for example, Mishan (1971), Schelling (1968), Freeman (1973). The methods discussed in these reports may be useful to the decision maker for comparative purposes, but none of the approaches is so convincing as to warrant adoption for regulatory purposes.

$60,000. Because he had a chance to save five additional lives at a cost of $60,000 each and chose not to, he reveals what is to him an upper limit on the implicit value of human life. Different values for life would be implied by different choices. For example, the choice of a total ban would imply a value for life of at least $60,000.

REDUCTION TO DOLLAR VALUES

The primary way in which analysis can simplify decisions is by reducing the relevant factors to numbers that can be added, subtracted, and compared. The only unit generally considered feasible for doing this is the dollar, but the use of dollar values poses a great many problems.

Many factors cannot be satisfactorily expressed in dollar terms because they involve important values on which there is no agreement. We saw an example of this in the case of human lives cited above. Some factors, such as aesthetics or competition, are difficult or impossible to express in dollar terms, even apart from the value question, because they are not traded in the market and thus are not generally expressed or thought about in monetary units.

Welfare economics and traditional benefit–cost analysis are based on the assumption that individual welfare is of prime importance, that individuals are their own best judges in matters concerning their welfare, and that therefore only those values that result from the voluntary actions of individuals (e.g., market prices) should be used in calculating welfare measures. The argument for relying on free markets to allocate resources is based on the assumption that markets reflect individual values; but the very existence of government regulation denies this assumption and implies that imperfect markets do not promote welfare to the desired degree. The government regulates chemicals because market prices do not reflect the full costs of chemicals to individuals. Value judgments as to what constitutes welfare may play a more important part in a decision than will market prices.

A substantial portion of the effects of regulatory decisions can be described and measured in terms of changes in the availability of goods and services to individuals. The analytical methods of economic theory and applied welfare economics can be used for the purpose of estimating dollar values which reflect opportunity costs and an individual's willingness to pay. Under certain assumptions market prices can be used directly for valuing goods and services or for assigning prices to such things as recreation or pollution, which do not themselves pass through markets. These assumptions are that there are no external economies or diseconomies or other forms of market failure, and that the existing distribution of income and purchasing power in the society is optimal. There is almost no situation for which these assumptions hold true, although for some aspects of decision making they may be sufficiently valid to permit the use of market prices as a first approximation of the values involved.

A final and major difficulty with using market prices is that the definition of benefits and costs differs, depending on the vantage point adopted. The benefits and costs for a particular community or state or income group will be different from the benefits and costs for the nation considered as a whole. This difficulty arises from the broader problem of the distribution of benefits and costs.

DISTRIBUTION

The way benefits and costs are distributed through time and across different segments of the population is crucial in many regulatory decisions, but quantification, conversion to dollar terms, and aggregation often obscure important distributional considerations. It is difficult to compare benefits and costs from multiple vantage points, such as the viewpoint of different regions of the country or different occupational classes. The function of the government decision maker is to weigh the impact of a decision on numerous groups in society; so it follows that even when the benefits or costs can all be stated in common dollar units, they may not be additive. From the vantage point of society as a whole (and traditional economics) the closing of a plant that employs 1,000 workers is not a cost, because, to the extent that there is full employment in the society, the workers will find employment elsewhere, and the productive value of the plant will similarly be put to work elsewhere in the economy. From the vantage point of the workers and the community in which the plant is located, however, the cost may be great. (The costs to the community that loses the plant may, of course, be canceled out by benefits accruing to another community that gains the workers and capital that had been located in the losing community.) Yet the cost to the workers and the community cannot be added to other societal costs of the government action if the vantage points for calculating the costs are different.

The distribution of benefits and costs over time poses even greater difficulties. In traditional benefit–cost analysis, the value of future benefits and costs is reduced by using a discount rate. One justification for using a discount rate is that resources invested to yield future benefits cannot be used now for present benefits; thus the discount rate reflects the opportunity cost of postponing consumption or benefits, just as interest earned on savings accounts compensates for the postponement of spending. The use of discount rates has been the subject of much dispute in traditional benefit–cost analysis, but these disputes are simple compared to the problems that arise in considering the future impact of regulating chemicals.

In the context of regulating chemicals, compensation for postponed spending or lost income from investment can be considered to generally reflect a preference for current benefits over future ones. A benefit in the hand is worth two in the future. Another reason for discounting, one on which there is some disagreement among economists, is the uncertainty of the future. Scientific and technological advances may change completely the current considerations involved in a decision, as, for example, when a simple cure is found for a disease. The application of a discount rate based on these justifications can cause considerable difficulty. For example, if the discount rate were 5 percent, 100 cases of poisoning 75 years from now would be equivalent to about 3 cases today; or 1 case today would be valued the same as 1,730 cases occurring in 200 years, or the same as the current world population (more than 3 billion cases) in 450 years. Clearly, intergenerational effects of these magnitudes are ethically unacceptable, yet they might be made to appear acceptable if the traditional social rate of discount concept were applied. There is as yet no generally accepted method for weighing the intergenerational incidence of benefits and costs.[10]

UNCERTAINTY

A characteristic of most decisions to regulate chemicals is that the full consequences of the decision are impossible to predict with any certainty at the time the decision is made. For example, the effect of a regulatory control on the distribution of a chemical in the environment, the exposure of people to the chemical, and the relationship between exposure and effects on health are often not well understood. Also, the economic data used to analyze economic impacts are typically old (i.e., not representative of the immediate situation) by the time they are

[10]For a discussion of this problem and a proposal for its solution, see Page (In press).

processed into usable form, and their reliability declines rapidly thereafter. At best, the data will be imprecise, and the cost and benefit analyses that result should be considered as probabilistic estimates of the magnitudes under examination. The use of uncertain data adds a significant subjective element to any analysis.

One of the contributions of decision theory is to show how probabilities can be used to quantify uncertainties. People are accustomed to hearing probabilities of rain in predictions of tomorrow's weather or odds on forthcoming elections and sporting events. The same concepts may be used to describe uncertainties related to toxic substances. An example of such a probability statement would be: "There is a 90 percent probability that the chemical will produce no increase in the incidence of cancer, and a probability of 5 percent is assigned that the chemical is weakly carcinogenic and would result in an annual increase of 1 to 100 additional deaths from cancer. A 5 percent probability is assessed that the compound is strongly carcinogenic in man and at least 100 deaths from cancer would result from each year of unrestricted use."[11]

Given the above limitations on benefit–cost analysis, there is no substitute for good judgment on the part of the decision maker.[12] Progress toward overcoming these limits of analysis is likely to be slow and uncertain, and the most important limitations are unlikely ever to be overcome.

Highly formalized methods of benefit–cost analysis can seldom be used for making decisions about regulating chemicals in the environment. Thus the development of such methods should not have high priority. However, the benefit–cost and decision frameworks described in this report can be useful in organizing and summarizing relevant data on regulatory alternatives which the decision maker must review.

THE USES OF ANALYSIS

We began with a discussion of the limitations of benefit–cost analysis to dispel exaggerated ideas about what such analysis can do, and to emphasize the types of problems that surround any attempt at formal analysis of costs and benefits. But we should remember that the relevant baseline against which analytical methods must be judged is not some

[11]The use of this type of probability statement is based on acceptance of what is known as the Bayesian approach to probability. Many statisticians do not accept this approach, but it is an essential element in decision theory.

[12]We recognize that good judgment also presents its problems. See Steinbruner (1974).

ideal system of value-free mathematics, but rather the intuitive, ad hoc, unsystematic, and often frantic methods now used. Compared to the latter, methods of benefit–cost analysis, using techniques and concepts adopted from economics and decision theory, represent a significant advance.

The method of analysis that we propose is described more fully in the Working Paper on Hazard–Cost–Benefit Comparison (Appendix H). The method can help to ensure an orderly examination of relevant factors and significant issues involved in the decision. It cannot in any sense be relied upon to determine what the decision should be, but it provides a systematic framework for reviewing the important elements of a decision. In short, it allows for decisions to be made by informed judgment rather than by uninformed intuition.

The following steps, in the order given, constitute a procedure for using the proposed analytical framework.

1. Specify the decision or problem of choice to be addressed by the analysis. The aim of the analysis is to assist the decision maker in making a choice. The question to be answered must be clearly stated so that the analysis can be focused on defining and evaluating alternative answers.

2. Review the available information on the past and present justification for regulating the chemical of interest. This step may involve nothing more than specifying the existing set of standards and regulations, or it may involve a detailed chronology of the development of standards and guidelines for this and all analogous materials. The information collected in this step should establish a base level of current information and data upon which the following steps can be built.

3. Describe the flow of the chemical from production through refinement and use to eventual disposal. The description should be quantified to show the materials balance from stage to stage and in particular the escape of emissions into the environment at each stage.

4. Trace the flows of the chemical from point of emission into the various media of the environment that lead to the exposure of people or other important components of the environment. Environmental monitoring data will play a crucial role in this step of the analysis. Also at this step uncertainty modeling should be introduced. No model will account with complete accuracy for the path and amounts of a chemical from emissions and dispersion to the points where it creates an exposure hazard, so probability statements or confidence limits need to be placed around the resulting estimates.

5. Estimate the effects of exposures on people and the environment.

Effects on human health can be described by means of a dose–damage function with confidence limits to describe a band of probable likelihood of damage.

6. Estimate the costs and benefits of alternative controls on the chemical, and project the economic impacts of imposing the alternative controls on the chemical production system.

The costs and benefits of alternative regulatory options must be compared with the alternative of no new regulation. Also, the estimation of the effects of imposing a regulation must take into account the substitute products or processes that would come into use as a result of the regulation. In other words, substitute products and processes must be an integral part of analyzing the costs and benefits of regulatory options. The market effects of the existence of substitutes are incorporated in the price, supply, and demand of the chemical under consideration. But the hazard and other nonmarket effects of substitutes must be separately considered. A decision on whether to ban DDT could not be made rationally without analyzing the hazards of parathion and other substitute pesticides. A decision to discourage the use of phosphates in detergents should not be made without comparing the existing situation to the costs and benefits of the new detergent formulations that would be used in place of phosphates. The costs of imposing a regulation on a chemical include the costs of the substitute chemicals that would come into use, and the same is true of benefits.

The results of the analysis must be summarized and arranged for presentation to the decision maker. A suggested format for summary display is shown in Figure 2. Without attempting to be inclusive, this display matrix shows how such a summary might be arranged. The comments column permits the analyst to call attention to those considerations that may be crucial in any particular case, such as degrees of uncertainty that may affect the final outcome.

It may be useful to extract the most important elements of the analysis for visual display. For example, Figure 3 shows the impact of three alternative control programs in terms of both expected values and uncertainties. Both hazard and benefit changes are represented as incremental changes from the current situation, so they will both presumably be reductions rather than positive values. Confidence limits are shown to reflect uncertainties in the estimation of the effects of the decision.

If all nonhealth economic costs and effects are combined in money terms on one dimension and hazards are shown in terms of expected

	COMMENTS	ALTERNATIVES			
		No Regulation	Ban Now	Control Option A	Control Option B
I. HAZARDS AVOIDED					
a. Health					
b. Environmental					
II. COSTS OF CONTROL					
a. Direct					
b. Indirect					
c. Market Structure					
III. BENEFITS LOST					
IV. DISTRIBUTION OF BENEFITS AND COST					

FIGURE 2 Display of major benefits and costs of regulatory alternatives.

deaths avoided on the second dimension, as they are in Figure 3, value judgments about the trade-off between the two can be represented by a family of lines called indifference lines.[13] Each line represents combinations of reduced hazards and net economic costs; concerning these combinations the decision maker is indifferent. An indifference line is analogous to a contour line on a map showing heights above sea level. Through any point on the chart there is an indifference line, and each

[13]In many types of analysis indifference *curves* rather than lines are used.

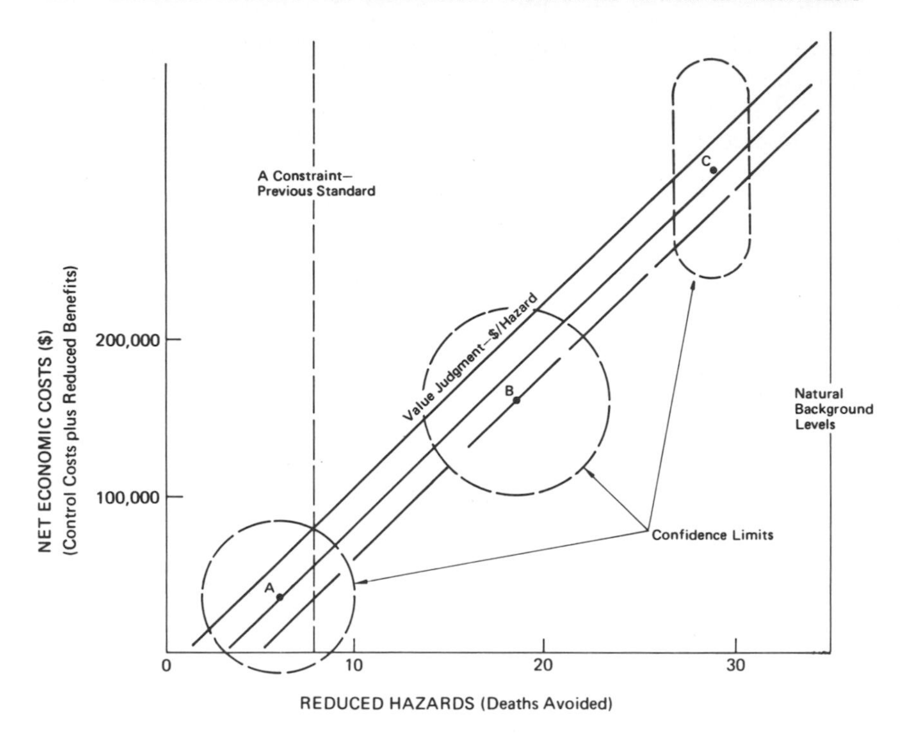

FIGURE 3 Value judgments in the hazard–cost–benefit display. See text for explanation.

line represents a different level of welfare as perceived by the decision maker. The objective of the decision maker is to choose that alternative regulatory option which lies on the indifference line closest to the lower right-hand corner of the figure. This would represent the highest obtainable level of welfare, because lines below and to the right represent both fewer deaths and smaller reductions in benefits. The slope of each indifference line is determined by the willingness of the decision maker to trade off net economic costs in order to achieve reduced hazards.

In Figure 3, Alternative B is preferred given the value judgment concerning values of deaths avoided. However, the implications of a higher value for deaths avoided can easily be portrayed. A higher dollar value means a more steeply sloped family of indifference lines. At least for some higher dollar values, Alternative C would be preferred.

The background level of pollution and its associated hazards to health and the environment represent a lower level of hazard reduction which cannot be surpassed. This constraint (maximum attainable reduction in hazard) is shown as a vertical line in Figure 3. Sometimes an upper level of permissible hazard may also be introduced because society is not willing (either currently or in some long-term sense) to accept a group hazard greater than some level. Preexisting standards or accepted guidelines may represent such a constraint.

To reflect distribution considerations (i.e., who gets what benefits and who accepts what hazards), alternatives shown on the hazard–cost–benefit chart can be duplicated for different interest groups. For example, one trade-off might represent Alternative A for a particular local area, while a different trade-off represented Alternative A for the United States as a whole. When interest-group considerations are shown in this way, the decision maker can make his own judgment of the appropriate weightings for the two trade-offs.

The chart assumes that the elements of the decision can be aggregated into only two components—benefits and hazards. In many cases this may not be possible, and even when it is possible, such aggregation entails a variety of assumptions about the weight and value of different elements. However, the matrix display of the type shown in Figure 2 should be sufficient to prevent the decision maker from being trapped into accepting unspecified assumptions.

As a final step, the same kind of analysis and graphical display can be used to indicate the types of information needed to most effectively narrow the uncertainties for future decisions ("feed-forward" information).[14] This step provides a systematic method of planning for future data integration at the same time decisions are being made on the basis of inadequate current data.

The approach outlined above deals with the problems of values by trying to make explicit to the decision maker what values are implied by various choices among the regulatory options. It may seem callous to trade dollars against human lives, but such valuation problems are unavoidable and are clearly implied by most of the important decisions the government makes. Our thesis is that since valuation of noncommensurables is unavoidable, it is better for the decision maker to confront the choice of values openly and explicitly than to allow values to be hidden. In actual decisions the trade-offs are unlikely to be as brutally straightforward as dollars against lives, because there will be several incommensurable elements, not just one.

The problem of reducing numerous elements in the decision to a few aggregated values is the major obstacle to full implementation of the

[14]A current analysis of sulfur oxide emissions (NAS 1975a:54–711) illustrates a method to assess the potential value of further information.

proposed analytical framework. The next two chapters are devoted largely to a discussion of this problem. But it should be kept in mind that the main contribution of the decision framework described here is not to make the decision or decide the outcome, but to organize information for the decision maker to assist him in the unavoidable task of balancing incommensurables and exercising judgment. The approach permits information from various specialists to be introduced unambiguously into the decision process. It permits the decision process to be reviewed by the various parties concerned; and where disagreements exist on the preferred decision alternative, the framework will help to reveal the basis of the disagreement. Furthermore, as new information is obtained, the decision can be updated rapidly and efficiently by incorporating the new information into an existing analytical framework. But weighing uncertainties and making trade-offs must in the end be done by the decision maker himself.

Value judgments about noncommensurate factors in a decision such as life, health, aesthetics, and equity should be explicitly dealt with by the politically responsible decision makers and not hidden in purportedly objective data and analysis.

7 Information on Hazards and Costs

DEFINITIONS AND HAZARD RATINGS

For the purposes of this study it was agreed that we would focus on the costs and benefits *of the chemical* as they would be affected by a proposed regulation. Thus, costs are defined as consisting of the following elements: *hazards* to man, the environment, and materials; *direct economic costs,* consisting of transaction costs, such as the research, monitoring, and enforcement costs of the regulation, and control costs, which are the immediate costs to industry, labor, and society of complying with a proposed regulation; *indirect economic costs* arising from the links between the regulated chemical and complementary or substitute chemicals and other products; and *market structure costs,* which include the potential effect of a regulation on concentration in the industry and on product innovation. The *benefits* of the compound subject to regulation include such elements as increased productivity, national security, aesthetics, and health.

Whether particular elements in a decision are considered costs or benefits depends on the vantage point used for analysis. If we had focused on the benefits and costs of the regulatory action instead of the chemical, the same elements would have been included in the analysis. However, the primary benefit would have been a reduction in hazard, and reduction in the benefits of the chemical would have become a major cost item.

The hazard of a chemical is a function of both the intrinsic toxicity of the compound and its use. Most classifications of chemical hazards have relied only on toxicity, but because the amount and type of use determine the extent to which man and the environment are exposed to the chemical, the use factor should be a basic component of any hazard rating scheme.[15]

There exist a variety of established methods for evaluating the toxicity portion of the hazard rating. These methods have been extensively discussed in the report *Principles for Evaluating Chemicals in the Environment* (NAS 1975b). However, the development of new concepts in human and environmental toxicological procedures must be continually encouraged.

Generation of the usage component of the hazard rating may be difficult. The usage factor must reflect temporal variation in exposure patterns, represent both the probability and intensity of exposures, and yet be reducible to a numerical scale in order to be readily usable. This formidable combination of conditions poses a challenging conceptual problem, but one which should be amenable to systematic solution. A reliable hazard rating system would be an important tool in aiding decision making on chemical regulation. The initial rating system probably would apply only to human health hazards. But hazards to nonhuman aspects of the environment are also of great importance, and thus future work would have to be devoted to methods for rating these other hazards.

The Department of Health, Education, and Welfare, in conjunction with EPA, should attempt to develop a hazard rating system, placing particular emphasis on evaluation of use patterns.

INFORMATION FOR DETERMINING HAZARDS

As already noted, *Principles for Evaluating Chemicals in the Environment* dealt comprehensively with the kind of information needed to evaluate health and environmental hazards. We emphasize here only a few additional points.

The first and fundamental point is our belief that society suffers large and unnecessary expense because of inadequate investment in determining the hazards of chemicals. The less precise the determination of hazard, the larger must be the necessary margin of safety incorporated in

[15]One such scheme that does consider both toxicity and use has been developed by the Stanford Research Institute and the National Cancer Institute. However, it was designed and is used for the purpose of establishing priorities for testing chemicals for carcinogenicity, not for regulatory purposes.

the regulatory standard to protect society. It usually will cost more to institute the more stringent controls resulting from the larger safety (ignorance) margin than it would to improve the precision of the data base.

The quality of chemical regulatory decisions is dependent largely upon the adequacy of the available information. To develop an adequate data base, research efforts in basic clinical and environmental toxicology and epidemiology and in economic analysis must be strengthened, and professional training in these areas must be supported. An interagency committee consisting of the relevant federal research and regulatory agencies should be established to maximize the use of existing information.

A major source of underutilized information is the large chemical manufacturers. The competitive aspects of business enterprise prevent the open exchange of product information between producers and governmental agencies. Equitable means of divulging such information must be devised. These could include the acceptance by the governmental agency of information from the producer on a confidential basis, the reimbursement by a competitor to a producer of a fair share of the cost of obtaining the information, or the joint funding by all producers of any information-seeking program. A major step toward joint funding by producers has recently been taken with the formation of the Chemical Industry Institute of Toxicology, an organization funded by the major chemical producers to test and evaluate potentially toxic chemicals.

The Department of Health, Education, and Welfare and EPA *should establish a task force, including, among others, representatives of the chemical industry and the scientific community, to develop a system for making unpublished or proprietary data about chemicals available to governmental agencies.*

Because of the large number of chemicals that require evaluation and assessment for regulatory decision making, the regulator usually cannot wait for the full development of all necessary information; thus the initial estimates of hazard tend to be less precise than desirable. The hazard estimates should be periodically reviewed regardless of the presumed quality of the original information. It thus is necessary that an appropriate feedback system for monitoring or surveillance be established to: (1) reevaluate and thus validate or invalidate the initial estimates; (2) assure that permissible concentrations that protect against unacceptable adverse effects are being maintained; and (3) check that the actual levels of exposure occurring in the community are not responsible for some new and unexpected effect. The only way to identify health hazards that have escaped detection by prescreening or other toxicological testing is by epidemiologic or population surveillance studies.

EPA, *in cooperation with the Department of Health, Education, and Welfare, should develop and use monitoring systems that can detect changing patterns in concentrations of specific toxic substances in biological tissues. They should also develop and use population surveillance systems that reflect changes in illness and death patterns due to environmental pollutant exposure. Data from monitoring systems and from other sources should be used to adjust past decisions when necessary.*

PRESENTATION OF HAZARD INFORMATION

Information on the health hazards of a chemical is too value- laden and too important a factor in the decision process to be aggregated with other aspects of the decision. However, the health data could usefully be summarized in a hazard rating when a reliable system for such ratings is developed. The effect of alternative regulatory options on the hazard rating would then summarize for the decision maker the effect on human health of different decisions. In any case, information on health effects should be accompanied by information on the probability that the effects will occur and also by an indication of the reliability of the data.[16]

The adverse health effects of a chemical do impose real monetary costs on society. Thus data such as the hospital costs for treating victims of disease related to a chemical may be useful to the decision maker. But such dollar measures are often inadequate indicators of the full impact of the hazards of a chemical.

Important adverse ecological effects should be treated in the same manner as health effects. However, other types of hazard information, such as damage to materials, might be aggregated into a dollar figure.

Aside from the information on hazards, there are three other types of costs that must be considered in a regulatory decision: (1) direct

[16]Probability, when applied to health effects data, can have two distinct but related meanings. One is the probability that a given individual in the population will suffer from the effect. Thus to return to an example used earlier, the statement that ". . . a 5 percent probability is assessed that the compound is strongly carcinogenic in man and at least 100 deaths from cancer would result . . ." means in part that if the compound is strongly carcinogenic, the probability that an individual will get cancer from the chemical is approximately 100 in 200,000,000 or .00005 percent, assuming that the total U.S. population is equally exposed. The other meaning is the probability that the data for the estimate are correct. This meaning is represented by the 5 percent figure in the quoted example.

economic costs, (2) indirect economic costs, and (3) market structure costs.

DIRECT ECONOMIC COSTS

Regulation imposes economic costs on society to the extent that conforming to the regulation absorbs resources (land, labor, capital, and so on). These costs are best measured by the value of these resources in alternative uses, i.e., by their opportunity costs. When markets for goods exist and are perfectly competitive, the market prices of these resources are the appropriate measure of their opportunity costs. When market prices are distorted by externalities, market failures, or monopoly elements, then the opportunity costs must be estimated from whatever relevant data are available.

Direct economic costs may be conceived as falling into two general types, transaction costs and control costs. Transaction costs are administrative costs incurred, usually by the regulating agency, in the process of collecting information and enforcing a regulatory decision. The extent of additional transaction costs is, of course, a function of the form of regulatory action taken. Great variation in transaction costs may exist depending on the regulatory option chosen, and their magnitude may itself be a factor in selecting the appropriate regulatory action.

Control, or what might also be termed compliance, costs consist of a more complex set of direct cost effects. In general, these are the impacts directly related to the production and distribution of the substance at issue. These can frequently be most easily perceived as changes in output, but even here, the nature of the production process involved requires examination of the possible impact on jointly produced products and on by-products.

It is obvious that control costs can be minimized when the decision to regulate is made early in the sequence between innovation and introduction of a chemical. This is also true of the indirect economic costs discussed below.

INDIRECT ECONOMIC COSTS

Indirect economic costs are considerably more difficult to assess, and attempts to include them in the analytical framework should be preceded by a judgment that the apparent balance between costs and benefits is likely to be altered by their inclusion. In general, these indirect economic costs arise from the commonly observed phenomenon of linkage among economic units.

It may be useful to visualize these linkages as arising first from the fact that products are frequently either complements to or substitutes for each other. In the complementary case, use of one substance makes likely the use of another so that regulation of one may have notable effects on the use of the other. Alternatively, when substitution is feasible, regulation of one may radically alter the use of the other. In both cases, there may be hazards induced through a change in use that may require a complete cost analysis.

Another variety of linkage exists because firms or plants are customers or suppliers of each other. The regulation of a product conceivably might have a negligible direct impact, but it may be a critical input to suppliers of other products.

These effects, which are a route by which regulatory impacts spread through the economy, are also reflected in changes in trade relations. The volume and direction of trade internally and with foreign countries can obviously be altered if existing buyer–seller relationships are altered by regulatory decisions. To assess fully the costs that may be associated with these decisions, it may be necessary to estimate not only the adjustments in production inputs and outputs that could result, but also what those adjustments imply for trade among geographic areas. There is, therefore, a regulatory need for detailed information about trade in toxic substances for use in benefit and cost assessments.

There is considerable danger of double-counting when indirect costs are included in the benefit–cost analysis. For example, when the chemical to be regulated is used as a component of other products, this will be reflected in the benefit measures of the chemical and should not be included again as an indirect cost. In general, it is only the effects that are not reflected in direct market transactions that should be included as indirect costs.

MARKET STRUCTURE COSTS

Market structure costs include the effect of regulations on the number of firms that compose the chemical industry, on the extent to which the production of given products is concentrated or dispersed among firms, and on the search for new products and processes within the industry.

These costs are important, but they are difficult to incorporate into the decision-making process for at least two reasons. First, they cannot be expressed in dollar terms that are compatible with other dollar costs. Second, regulatory decisions may have cumulative impacts on structural costs that are difficult to associate with any single decision.

PRESENTATION OF COST INFORMATION

Both direct and indirect economic costs can be expressed in dollar terms derived from generally available economic data. However, the distributional effects of indirect economic costs will require analyses that may be expressed in dollar terms but which are not capable of being added or otherwise combined with other dollar costs. Structural costs also can be described in quantitative terms but not in terms that can be combined with the other types of cost. For example, the impact of regulations could be expressed as the effect on the five largest producers' percentage of total industry output.

Many costs are not subject to direct measurement. Consequently, cost data must be viewed as partially judgmental at best, and even where direct measurements are possible, they are often imprecise and outdated. This is not as great a handicap as might first appear once it is recognized that regulation is not a mathematical procedure. Cost information can be adequate to provide the regulator with an understanding of the relative magnitudes involved and of the form and direction of probable adverse changes. This, however, requires conceptual skill and investment of resources to specify the information required and the costs and benefits of obtaining it.

8 Information on Benefits

IMPORTANCE OF BENEFITS

The use of chemicals provides an enormous variety of benefits.[17] Chemical products such as soap, chlorine, drugs, and vitamins have enhanced and extended human life and reduced suffering and pain. Human work effort has been reduced, national security increased, and natural resources conserved by chemicals. Both convenience and aesthetic pleasure are derived from a variety of chemical products.

The impact of the benefits of a chemical is often diffuse, both in the sense of affecting a large number of people and in the sense of not exerting an impact on any single individual or institution in easily identifiable or quantifiable terms. Practically speaking, in many cases only the manufacturer and his employees are likely to focus on the foregone benefits of a proposed regulatory action, whereas the government and the general public are more concerned about hazards and costs. As Rita Campbell (1974) has said

> A difficulty in weighing large risks, sometimes of unknown probability, to a few persons against relatively small benefits to many people is that the significance of the large risk and its possible horror to the individual are more easily comprehended and thus well publicized. On the other hand, the relatively small benefit to each individual has little public impact, even though when multiplied manifold it may become much larger in total. Thus even the existence of these benefits, actual or potential, is often unknown by the person who may benefit.

[17]For a full account from the viewpoint of industry, see American Chemical Society (1973).

Because of these factors, benefits may receive less weight than they deserve in regulatory decision making. However, benefits are sometimes exaggerated because of a failure to recognize the availability of substitute products or practices which can be used if regulatory action limits the use of a particular chemical.

METHODS FOR MEASURING BENEFITS

If some form of regulation is imposed on a chemical, it is likely to affect the availability of the chemical through changes in its price and/or the quantity of the chemical produced; and this, in turn, means changes in benefits and welfare. The question is how we can measure the changes in welfare associated with changes in the prices and quantities of the goods and services associated with a regulated chemical.

Economists are clearly unable to measure welfare. It has many dimensions and is affected by many forces both within and outside the sphere of economics. However, economic analysis can be used to identify and measure one component of welfare, that part of it which stems from the availability of goods and services produced by people using scarce resources—land, labor, and capital. Although this is not the only component of welfare, it is likely to be an important component in regulating chemicals.

The prices of chemicals that appear in the market summarize better than any other indicators the sum of expected net benefits as judged by purchasers. Since most purchases are part of an ongoing repetitive purchase pattern, rather than once-in-a-lifetime purchases, there is a presumption that most purchasers received the benefits they expected. Those who were disappointed ceased to buy, even as new purchasers were first trying the product. Price measures the marginal benefits buyers expected from those purchases, if we assume that the consumers were as aware of private hazards and costs as they were of private benefits, and that there were no important social costs or benefits overlooked in the private decision process.

These judgments are peculiar to time and place. If consumer incomes were different (or prices, or knowledge, or government regulations, or similar factors), there would undoubtedly have been different expenditures. But that succession of "ifs" does not create that alternative world. Given the prevailing conditions of our society, price generally measures benefits expected by those who buy.

The very existence of government agencies charged with regulating chemicals indicates that Congress and the Executive are no longer prepared to take the private market judgment of chemical purchasers as a sufficient indication of the net benefits or costs of chemicals. The government must consider benefits in addition to those that private consumers recognize. As with costs, some benefits may have been noted by consumers but not recognized or reasonably weighed by them. Still other benefits, such as national defense or aesthetics, may not interest individual buyers at all though they may be important to society. The government's mandate, however, does not permit it to ignore any class of benefits to which any group of serious spokesmen wish to call attention. Thus it is likely that some portion of the benefits of a chemical will not be measurable in economic terms and will have to be separately considered by the decision maker.

Using the price of a product as a measure of benefit received poses some other major difficulties. Not the least of these is the fact that consumers might actually be willing to pay much more for a product than they actually do pay. The price consumers pay for water, for example, is not a good measure of the benefits they receive from it. To surmount this difficulty, many economists have urged the use of "consumer surplus" as a way of measuring the benefits of a product.

Three basic principles of applied welfare economics provide the basis for using consumer surplus to measure the net economic benefits of a change in resource allocation:

1. the competitive demand price for a given unit of a good measures the value to consumers, or their marginal willingness to pay;
2. the competitive supply price for a given unit of a good measures the value to producers, or opportunity cost;
3. the net economic benefit of an additional unit of a good is the value to consumers less the cost or value to producers (Harberger 1971).

An application of these principles is shown in Figure 4, in which the competitive demand and supply curves are drawn. The consumer's willingness to pay for the unit represented by *A* is represented by the vertical distance *AD*, while the cost of that unit is *AC*. The net benefit is *DC*. Furthermore, by the same reasoning, the net benefit of increasing the quantity available by *AB* units is the shaded area *CDE*.

There are many unresolved issues concerning the interpretation of consumer surplus and its relationship to individual utility and welfare, but the general concept is widely accepted in welfare economics and has been used as a basis for measuring the economic effects of public policy changes in such diverse areas as water resources development, pollution control, transportation investment, and the regulation of drugs.

Aside from some theoretical problems with the consumer surplus

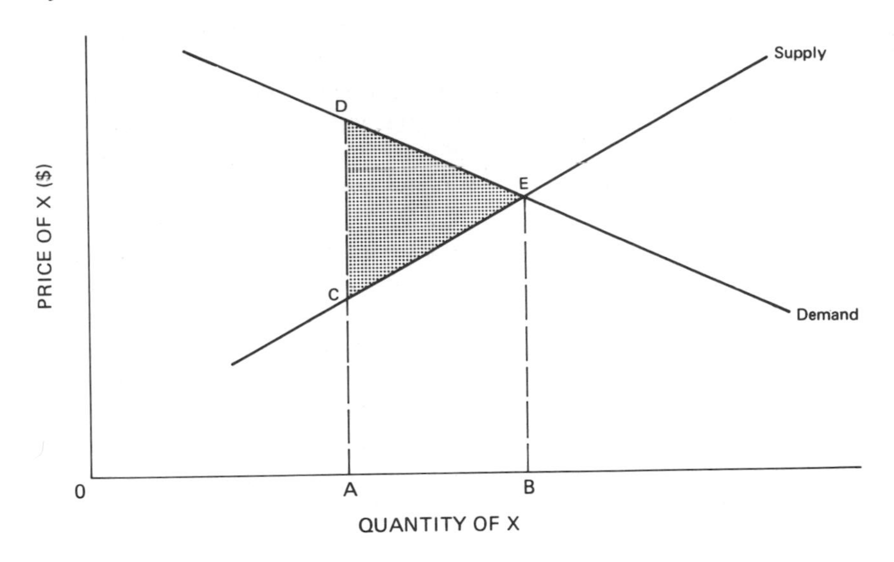

FIGURE 4 The consumer surplus measurement of net benefits. See text for explanation.

concept, it may be difficult or impossible in many cases to measure the consumer's "willingness to pay" (the demand curve in Figure 4). It is especially difficult to measure "willingness to pay" in situations where the change in demand is not marginal, such as the change that would be created by a ban on some uses of a chemical. In these situations, the type of analysis represented in Figure 4 may not be applicable.

In some cases, the availability of time and the value of the information may justify a major effort at econometric estimation of the demand curve. In all cases of existing chemicals, one point on the demand curve is known—current price and quantity; and in the absence of more precise information it may be useful to use expert judgment, taking into account such factors as available substitutes, to provide a range of estimates on the nature of the demand curve. Such judgments should be stated in probability terms to reflect their uncertainty as discussed in Chapter 6. Finally, there are short-cut techniques which can be used to provide consistent approximations of benefit measures in some fairly common situations. These techniques include net productivity, damages avoided, and cost of alternatives (Freeman 1973).

Regardless of what technique of measurement is used, it may be presumed that a private firm seeking permission to market a new product, or desiring to continue marketing an old one, is ready, willing, and able to provide the most comprehensive list of the classes of benefits which that chemical will provide. And by also providing sales figures, by type of user or application, the marketer will give an indication of the relative importance of the chemical. Data on the anticipated benefits of new chemicals will obviously be more speculative than data on the benefits of existing chemicals. But as a first approximation, the producer can reasonably be asked to supply what most companies today create before marketing a new product—namely, a market projection (with basic estimates of the price they expect to charge and the sales they expect over the first few years). The data may not be precise or wholly objective, but they constitute appropriate raw material for the regulatory agency, which can then subject them to tests of consistency and reasonableness.

These data would provide much of the basic information needed to analyze the benefits of a chemical in a regulatory context. But in many cases information would also be needed on the distributional aspects of the benefits—what groups benefit, where, and over what time period.

Once this potential complexity is recognized, the need for a method of displaying the benefits so that trade-offs can be visualized becomes obvious. The form of this display and the sources of the required data are important issues.

Generalization on these issues is difficult beyond stipulating the need for that categorization of group, place, and time which the facts in each case dictate. The sources for the facts have already been mentioned. Who will benefit, where, and when should be revealed by the same information on materials balances and economic flows that have been suggested previously as necessary elements in the decision process. From these it should be possible to deduce the nature and magnitude of the benefits resulting from alternative patterns of regulation.

CONCLUSION

Economic costs and benefits are likely to be basic considerations in regulatory decisions. Thus it is important that an appropriate conceptual framework be used to define and, where possible, to measure economic costs and benefits. When regulatory decisions are expected to cause changes in the output of goods and services, the net economic value of these changes should be measured in dollar terms, using the principles of willingness to pay and opportunity cost, and they should be included in the information presented to the decision maker.

All information on benefits and costs is subject to the many problems of value, equity, and so forth discussed at the beginning of Chapter 6. There is no fully objective way of arraying such information. But the costs and benefits can be measured approximately, and the resulting information can be aggregated into a sufficiently small number of elements that a decision maker can grasp the potential effects of a decision and can comprehend the differences in effect produced by alternative choices. The decision process can be made manageable while using the available data and incorporating the major elements that must be considered. We have not described a scientific way to make decisions, because we believe that to be impossible. But we have described a method of obtaining and handling information that we believe can improve the rationality of the decision-making process and that will thereby result in better decisions.

References

American Chemical Society (1973) Chemistry in the Economy. Washington, D.C.: American Chemical Society.

Campbell, R. R. (1974) Food Safety Regulation. Washington, D.C.: American Enterprise Institute for Public Policy Research and Stanford University: Hoover Institution, p. 22.

Dominick, D. D. (1974) HCB Tolerances: An Example of Risk-Benefit Decision Making. Available on request from the National Research Council.

Dorfman, R., ed. (1965) Measuring Benefits of Government Investments, papers presented at a conference, November 1963. Washington, D.C.: Brookings Institution: Studies of Government Finance.

Eckstein, O. (1958) Water Resource Development: The Economics of Project Evaluation. Cambridge: Harvard University Press.

Freeman, A. M., III (1973) A Survey of the Techniques for Measuring the Benefits of Water Quality Improvement, prepared for the EPA Symposium on Cost–Benefit Analysis in Water Pollution Control.

Harberger, A. C. (1971) Three Basic Postulates for Applied Welfare Economics: An Interpretive Essay. Journal of Economic Literature. 9(3): 785–797.

Hitch, C. J., and R. N. McKean (1965) The Economics of Defense in the Nuclear Age. Cambridge: Harvard University Press.

Mack, R. (1971) Planning on Uncertainty. New York: Wiley Interscience.

Mishan, E. J.(1971) Evaluation of Life and Limb: A Theoretical Approach. Journal of Political Economy. 79(4): 687–705.

National Academy of Engineering (1972) Committee on Public Engineering Policy. Perspectives on Benefit–Risk Decision Making. Washington, D.C.: National Academy of Sciences.

National Academy of Sciences (1974) Planning for Environmental Indices. Available from National Technical Information Service as PB 240971.

National Academy of Sciences (1975a) Analysis of Alternative Emissions Control

Strategies. In Air Quality and Stationary Source Emission Control. Washington, D.C.: U.S. Government Printing Office.

National Academy of Sciences (1975b) Principles for Evaluating Chemicals in the Environment. Washington, D.C.: National Academy of Sciences.

National Academy of Sciences (1975c) Conference on Air Quality and Automobile Emissions. Internal report by the National Research Council to the Environmental Studies Board Committee on Environmental Decision Making. Mimeo.

North, D. W. (1968) A Tutorial Introduction to Decision Theory. IEEE Transactions on Systems and Cybernetics, SSC-4 (3): 200–210.

Office of Science and Technology-Council on Environmental Quality (1972) Report of OST–CEQ Ad Hoc Committee on Environmental Health Research. Mimeo.

Page, T. (In press) The Economics of a Throwaway Society. Baltimore: Johns Hopkins University Press, for Resources for the Future.

Pratt, J. W., H. Raiffa, and R. Schlaifer (1965) Introduction to Statistical Decision Theory. New York: McGraw-Hill.

Schelling, T. C. (1968) The Life You Save May Be Your Own. In Problems in Public Expenditure Analysis, edited by Samuel B. Chase. Washington, D.C.

Steinbruner, J. D. (1974) The Cybernetic Theory of Decisions. Princeton: Princeton University Press.

Train, R. E. (1975) Remarks by the Honorable Russell E. Train, Administrator, EPA, prepared for delivery before the National Wildlife Federation. Pittsburgh. March 15.

Tribus, M. (1972) Rational Descriptions, Decisions, and Designs. New York: Pergamon Press.

APPENDIX A

List of Participants in the Working Conference on Principles of Decision Making for Regulating Chemicals in the Environment

J. CLARENCE DAVIES, III, Resources for the Future, Washington, D.C., *Chairman.*

NORTON NELSON, New York University, New York, New York.

PANEL MEMBERS

PANEL 1 ON GOVERNMENTAL OBJECTIVES

RICHARD FAIRBANKS, Ruckelshaus, Beveridge, Fairbanks & Diamond, Washington, D.C., *Chairman.*

WILLIAM BUTLER, Environmental Defense Fund, Washington, D.C.

MARSHALL MILLER, Reaves, Pogue, Neal and Rose, Washington, D.C.

HARRISON WELLFORD, U.S. Senate Committee on Commerce.

PANEL 2 ON THE ROLE OF NONFEDERAL PARTICIPANTS IN THE REGULATORY PROCESS: THE PUBLIC OR PUBLICS

JESSE L. STEINFELD, University of California, Irvine, California, *Chairman.*

GERSHON FISHBEIN, Environmental Health Letter, Occupational Health and Safety Letter, Washington, D.C.

PAUL KOTIN, Johns Mansville Inc., Denver, Colorado.

CHANNING LUSHBOUGH, The Rockefeller University, New York, New York.

GLENN PAULSON, New Jersey Department of Environmental Protection.
SHELDON W. SAMUELS, Industrial Union Department of AFL–CIO, Washington, D.C.

PANEL 3 ON EQUITY

WILLIAM A. THOMAS, American Bar Foundation, Chicago, Illinois, *Chairman.*
NICHOLAS A. ASHFORD, Massachusetts Institute of Technology, Cambridge, Massachusetts.
EDWARD BURGER, National Science Foundation, Washington, D.C.
TALBOT PAGE, Resources for the Future, Washington, D.C.

PANEL 4 ON GOVERNMENTAL INFORMATION NEEDS: BENEFITS

STANLEY LEBERGOTT, Wesleyan University, Middletown, Connecticut, *Chairman.*
ROY E. ALBERT, New York University, New York, New York.
BERTRAM D. DINMAN, Aluminum Company of America, Pittsburgh, Pennsylvania.
ROGER WEISS, University of Chicago, Chicago, Illinois.

PANEL 5 ON REGULATORY INFORMATION NEEDS: HAZARDS AND COSTS

DAVID P. RALL, National Institute of Environmental Health Sciences, Research Triangle Park, North Carolina, *Chairman.*
SEYMOUR FRIESS, Naval Medical Research Institute, Bethesda, Maryland.
IVARS GUTMANIS, National Planning Association, Washington, D.C.
VAUN NEWILL, Exxon Corporation, Linden, New Jersey.
MONROE NEWMAN, Pennsylvania State University, University Park, Pennsylvania.
WILLIAM PAPAGEORGE, Monsanto Company, St. Louis, Missouri.

PANEL 6 HAZARD–COST–BENEFIT COMPARISON

A. MYRICK FREEMAN, III, Bowdoin College, Brunswick, Maine, *Chairman.*
RITA RICARDO CAMPBELL, Stanford University, Palo Alto, California.
KENDALL MOLL, Stanford Research Institute, Menlo Park, California.
WARREN MUIR, Council on Environmental Quality, Washington, D.C.
WARNER NORTH, Stanford Research Institute, Menlo Park, California.
RALPH D'ARGE, University of California, Riverside, California.

PANEL 7 ON REGULATORY OPTIONS

DAVID L. JACKSON, The Johns Hopkins University Hospital, Baltimore, Maryland, *Chairman.*

DAVID D. DOMINICK, Dominick and Corcoran, Washington, D.C.

SHELDON W. SAMUELS, Industrial Union Department of AFL–CIO, Washington, D.C.

PANEL 8 ON MARKET AND PRIVATE SECTOR DECISION MAKING

JAMES L. GODDARD, Ormont Drug and Chemical Company, Englewood, New Jersey, *Chairman.*

ARTHUR W. BUSCH, Southwest Research Institute, San Antonio, Texas.

JAMES D. HEAD, Dow Chemical Company, Midland, Michigan.

OBSERVERS*

Fred Abel, U.S. Environmental Protection Agency, Washington, D.C.

James Brodsky, Consumer Product Safety Commission, Washington, D.C.

Michael Brownlee, U.S. Senate Committee on Commerce, Washington, D.C.

John Buckley, U.S. Environmental Protection Agency, Washington, D.C.

Kerrigan Clough, U.S. Environmental Protection Agency, Washington, D.C.

Michael K. Glenn, U.S. Environmental Protection Agency, Washington, D.C.

Gordon Hueter, U.S. Environmental Protection Agency, Research Triangle Park, North Carolina.

Raphael Kasper, National Research Council, Washington, D.C.

Ralph Kennedy, CONSAD Research Corporation, Pittsburgh, Pennsylvania.

Victor Kimm, U.S. Environmental Protection Agency, Washington, D.C.

Alan Maguire, CONSAD Research Corporation, Pittsburgh, Pennsylvania.

Jerry Penno, Consumer Product Safety Commission, Washington, D.C.

*Several of the observers contributed to the work of individual panels. In this regard, the Panel on Governmental Objectives is grateful to Douglas Worf; the Panel on Hazard–Cost–Benefit Comparison is grateful to Fred Abel and Jerry Penno; and the Panel on Regulatory Options is grateful to John Buckley and Michael Glenn.

Glenn Schweitzer, U.S. Environmental Protection Agency, Washington, D.C.

Dennis Tihansky, U.S. Environmental Protection Agency, Washington, D.C.

Jay Turim, U.S. Environmental Protection Agency, Washington, D.C.

Irvin E. Wallen, U.S. Environmental Protection Agency, Washington, D.C.

Douglas Worf, U.S. Environmental Protection Agency, Research Triangle Park, North Carolina.

APPENDIX B

Toxic Chemicals and Regulatory Decision Making: Philosophy and Practicality

GLENN E. SCHWEITZER

Director, Office of Toxic Substances
U.S. Environmental Protection Agency

THE GENERAL FRAMEWORK

Integration of economic, energy, and environmental concerns. The need to avoid extremes and exhibit moderation in applying environmental controls. Balancing costs, risks, and benefits. Maintaining environmental objectives while adjusting the short-term timetables to accommodate economic problems. The inevitability of some level of risk in everything we do.

This rhetoric, characterizing the current climate surrounding the environmental movement, is appealing and seems easy to grasp. But what does it really mean as we address specific regulatory actions directed to toxic chemicals?

On a macro scale, we sometimes feel we are able to express the dollar costs of environmental control measures with some confidence in our estimates. Recently, drawing on the practices of the statisticians, we have begun to manipulate data on health effects to quantify the level of risk and the size of the population at risk. At the micro level, it often seems still easier to estimate the impact of regulatory actions, answering such questions as: What will a specific pollution control device cost in a specific plant? What will the resultant contaminant levels be in the stream or in the drinking water? Is this level below a "no effect" level?

Address to the Working Conference on Principles of Decision Making for Regulating Chemicals in the Environment, New Orleans, Louisiana, February 18–23, 1975.

Assessing the aggregate effect of micro impacts, however, involves far more than summing up the parts. For example, a risk factor of 1/1,000,000 appears different when viewed from the perspective of the 1 than from the perspective of the other 999,999. And the cumulative effect of regulatory actions, taken together with the uncertainty of future actions, has a dramatic effect on the whole climate of doing business. Many aspects of this effect are difficult to identify, let alone quantify. Guessing the current and future attitudes of regulatory agencies is becoming a way of life within many chemical companies as they reach decisions on the amount of resources to devote to research and development, the level of investments in introducing new products or continuing old products in the face of evidence suggesting possible environmental problems, and the pollution abatement conservatism to be built into the design of new plants.

TRENDS IN REGULATORY DECISION MAKING

The legislative history of toxic substances regulation is punctuated with inconsistencies as to the weight to be given costs, risks, and benefits in decision making. The most recent addition to the annals of environmental law, the Safe Drinking Water Act, addresses directly the perennial question of the weight to be given to public health considerations in determining enforceable contaminant levels. The act calls for both (a) health based contaminant levels—or health goals—and (b) enforceable contaminant levels which explicitly take into account technological feasibility and reasonableness. Similarly, the many versions of the proposed Toxic Substances Control Act have consistently highlighted the need to weigh all relevant factors, including economic and technological considerations, in reaching regulatory decisions. As you know, this philosophy is not expressed with similar clarity in certain sections of the Clean Air Act and the Federal Water Pollution Control Act (FWPCA).

Meanwhile, we are increasingly requiring more broadly based assessments of the impact of regulatory decisions—environmental assessments, economic assessments, and, recently, inflationary assessments. Usually the procedure is to postulate a numerical standard for a toxic chemical or a specific type of limitation on the use of the chemical, with the restriction designed to reduce environmental levels to the point that concerns over health or environmental damage disappear. Then an assessment is carried out to see if the favorable environmental impact from the restriction warrants the concomitant economic costs. If the costs are too high, the level of control is adjusted until an appropriate balance is reached. Unfortunately, to date this balancing act has been ad hoc at best, while at the same time our new authorities are calling for more systematic approaches to such balancing. As the judiciary becomes more deeply involved in toxic substances, it seems essential that these approaches improve both in form and in substance if regulations are to be sustained by the courts.

In our zeal to balance risks and benefits in some quantitative way, we frequently try to place numbers on the risks associated with very uncertain health data. More often than not, in the quantification process we drop the margin of error from our deliberations. This has been particularly true in the carcinogen area. Largely as a result of the work of statisticians beginning 10 or more years ago, there are now techniques for estimating risk factors. However, too often the lawyers and economists seize upon these numerical risk factors, frequently forgetting that these experimentally derived estimates may in fact have a very shaky relevance to the real world.

NEEDS OF REGULATORY DECISION MAKING IN THE NEAR FUTURE

Clearly, our goal is to improve the decision-making process, a process that is usually plagued with insufficient time and unbelievably soft data. Nevertheless, decisions must be made, decisions that we hope will be both sensible and defensible. We need to better use existing data—to better assemble what we have, to see more clearly what we do not have, to make reasonable extrapolations across the data gaps, and then to make judgments which best reflect the interests of society. If legal constraints inhibit judgments which best serve society, we should seek changes in the law.

We need to better orient our data collection and manipulation activities so that the information is arrayed in a way more relevant to decision making. I need not describe for you the difficulties in trying to derive the roots for numerical standards on the basis of a toxicological experiment which was designed to expand the state of the art of toxicology. Finally, we need to make better integration of data concerning a variety of considerations—health, persistence, control technology, economic dislocations. Is it possible to move away from past comparisons of apples with oranges toward a successfully integrated regulatory system?

THE SCOPE OF THIS CONFERENCE

This meeting is concerned principally with advancing the state of the art in reaching regulatory decisions on toxic chemicals which for one reason or another have already been identified as highly suspect in causing serious environmental or health problems. The suspicion may have arisen from recent epidemiological findings, toxicological experiments, or monitoring data, or it may reflect an accumulation of scientific or public concerns not necessarily tied to recent revelations. It may have resulted from deliberate efforts to search out problem chemicals. In any event, given a problem chemical, how can we more effectively decide whether regulatory action is in order, and if it is, how tight or how loose the regulation should be? We are particularly concerned with decision making which must be carried out in the near term in the absence of what purists—or perhaps even reasonable men—would call adequate time or data. There is really no option to delay the decision until better data are available. This does not mean that the interim decision is necessarily the final decision; indeed, I have not been involved in a single regulatory decision of any significance that has not included as part of the package the requirement to collect better data so that the decision can be reviewed and perhaps improved at a later date.

There are several concerns closely related to the regulatory decision which are not a principal focus of this meeting, although it is recognized that you cannot totally neglect these aspects. One area of considerable importance which is being explored in a variety of other forums concerns methods for selecting from the world of chemicals those chemicals or chemical activities which require priority regulatory attention. A second area receiving considerable attention within the U.S. Environmental Protection Agency (EPA) is the selection of the most appropriate legal authority for reducing identified problems. For example, what are the respective roles of Effluent Guidelines and Toxic Effluent Standards under FWPCA; when are New Source Performance Standards more appropriate than Hazardous Pollutant Standards under the Clean Air Act; and which sections of the pesticide laws can most appropriately be invoked to curtail certain problems? A third important aspect relates to the generation of test data, a topic that has already been explored in some detail by the National Academy of Sciences (1975).

At this session we would like to emphasize the problems of making decisions with less than optimal data. A secondary interest relates to the shaping of data collection activities, particularly in those instances when there is sufficient time to improve the data base before a regulatory decision is necessary.

In general, we are concerned with two types of regulations: first, product limitations, where the manufacture, use, distribution, or disposal of a chemical is regulated, as in the case of pesticides or chemicals under the Toxic Substances Control Act; and second, the determination of the acceptable levels for pollutants—expressed as numerical standards or as a control technology requirement—as in the case of hazardous air emissions, toxic water effluents, drinking water contaminant levels, pesticides tolerances, and the like. Further complicating the task is our special interest in multimedia contaminants.

EXAMPLES OF PRACTICAL REGULATORY PROBLEMS

Let me cite three regulatory areas that exemplify the types of concerns this conference might address. During the next several years, EPA will be involved in a variety of decisions in all these areas. There are, of course, other areas as well, particularly pesticides and air emissions problems.

REGULATORY ACTIONS UNDER THE PROPOSED TOXIC SUBSTANCES CONTROL ACT

The proposed Toxic Substances Control Act would give EPA the authority to prevent or limit production, distribution, or use of selected chemicals or require that they be labeled with instructions concerning use or disposal. The most recent versions of the law require that in reaching regulatory decisions the administrator consider:

- the effects of the substance on health and the magnitude of human exposure;
- the effects of the substance on the environment and the magnitude of environmental exposure; and
- the benefits of the substance for various uses and the availability of less hazardous substances.

After considering these factors, the administrator could impose regulatory action if he determines that there is an "unreasonable risk" to health or the environment.

The purpose of this conference is not to try to determine what constitutes an unreasonable risk, since the judgment must be made on a case-by-case basis. However, we are hopeful that you can assist in further clarifying the factors that should be weighed in reaching such a judgment and in setting the ground rules for arraying data so that the judgment will be as fully informed as possible. Again, it is important to keep in mind that the data which will be arrayed will inevitably be far from complete.

As one point of departure in our internal efforts to assemble data in an orderly fashion for determining the need for regulations under this proposed law, we have identified the following five steps:

- assessment of the effect of various levels of exposure of a chemical or its derivatives on man or the environment;
- assessment of the likely levels of exposure through material balance analyses;
- confirmation by monitoring of actual exposure levels;
- determination of the likely reduction in exposure levels and attendant effects through alternative regulatory action; and
- assessment of the economic and social costs of the alternative actions.

Under each of these categories we are attempting to develop checklists of items to be considered. I am sure that these data categories can be improved, but perhaps of equal importance is how the data from these categories are to be meshed.

NATIONAL PRIMARY DRINKING WATER REGULATIONS

In about two years EPA will propose revised National Primary Drinking Water Regulations for contaminants in drinking water, including probably several dozen chemicals. These regulations must take into account both public health considerations and feasibility of attainment. Thus, it seems necessary to consider (a) health effects at various contaminant levels, (b) monitoring data showing the current and projected range of contamination in water supplies, (c) costs involved in reducing high levels to lower levels at the water supply, (d) feasibility and costs of monitoring low levels, (e) sources of the contaminants, and (f) costs of reducing these sources as an alternative approach to cleaning up the water after it is contaminated.

The problem of data collection and analysis is particularly complicated, because a single standard will apply to 40,000 water supplies of all sizes and characters. Short-term variances will help with some of the extreme cases, but the need for reliable statistical sampling will remain a complicating factor. The problem is at least twofold: collecting and displaying the data in these categories in a meaningful way, and integrating the data in a way that will enhance the soundness of judgments as to required levels. Again I would add that much of the data may be soft and not susceptible to easy quantification.

TOXIC EFFLUENTS THAT CAUSE HEALTH PROBLEMS

We are now involved in a major effort to determine which chemicals currently being discharged into our waterways constitute a health hazard. One starting point is an examination of monitoring information from drinking water supplies, fish products, and recreational areas concerning the presence of chemicals, followed by isolation of those chemicals that are (a) suspected of having adverse effects on health at relatively low levels, (b) discharged by industrial facilities, and (c) sufficiently persistent to reach man via the recreational, drinking water, or food routes.

It is not difficult to identify a few such chemicals; however, it is more difficult to determine whether effluent controls on these chemicals arc warranted and, if so, the degree of control that is appropriate. The chemicals will usually be present in trace amounts, and the hazards to health from these trace amounts, even for carcinogens, are far from clear. The incremental decrease in the trace amounts, and the attendant health implications, which will be achieved by turning off or reducing selected sources is similarly not clear, particularly if industrial dischargers are not the only source of the contaminants. Finally, whether the control costs are warranted by the incremental health gain is also far from straightforward. We anticipate that we will be grappling with this problem for several years, and advice and suggestions on appropriate methodologies will be very welcome.

THE OUTPUT FROM THIS CONFERENCE

What can be realistically accomplished during the next several days and during the report writing that will continue for several months?

Perhaps most importantly, a greater sensitivity to the multiplicity of societal interests bearing on regulatory decision making will emerge—certainly among the participants and, it can be hoped, among the recipients of the report. This sensitivity should in turn lead to a higher level of sophistication in our decision making.

General principles which can serve as guideposts in the field of chemical regulation will be welcomed by all participants in the regulatory process—the regulator, the regulatee, and the adjudicator. Such principles are particularly timely as the influx of new people into this field continues to grow. The scope, general applicability, and specificity of these guideposts are the crux of the deliberations during the next several days.

With the benefit of hindsight, we should be able to identify common shortcomings in past decisions which an improved methodology can overcome. And surely as we consider the types of decisions ahead, specific ideas and suggestions will emerge.

Finally, we are searching for practical suggestions as to how we can significantly improve our data base for decision making without "breaking the bank."

THE CONTROL OF TOXIC SUBSTANCES IN THE YEARS AHEAD

As far as toxic substances are concerned, the quality of life in the chemical age of the 1980s depends, in large measure, on the outcome of the competition between economic growth, on the one hand, and the sophistication in approaches to responsible regulation—by both industry and government—on the other hand. Clearly, more people will be exposed to more chemicals in more situations. It is to be hoped that we can develop the necessary precautionary measures that will limit exposure to chemicals when necessary while not unnecessarily curtailing commercial activities.

There is a danger that society will not act responsibly in the matter of toxic substances through its government and other institutions. Endless legal confrontations will be the inevitable outcome, and the entire approach to toxic substances could bog down in the courts. To avoid such a situation, the highest quality of governmental leadership is essential—leadership characterized by technical credibility and openness in dealing with controversial data. You have a genuine opportunity to help stimulate such leadership.

The balancing of costs, risks, and benefits is not unique to the control of chemicals; it is the essence of many of our routine daily activities, as when we drive our car to the corner drugstore for a carton of cigarettes. However, when the beneficiary of an activity is different from the person exposed to its risk, we can no longer be casual in balancing costs, risks, and benefits. A more deliberate approach to societal decisions is the challenge of the years ahead.

REFERENCE

National Academy of Sciences (1975) Principles for Evaluating Chemicals in the Environment. Washington, D.C.: National Academy of Sciences.

APPENDIX
C

Working Paper on Governmental Objectives

SUMMARY

This paper on governmental goals in the regulation of chemicals (1) addresses the existing legislative framework as conceived by Congress and interpreted by administrative agencies and the courts; (2) analyzes briefly some of the factors affecting current administrative practices within the federal executive branch; (3) reviews the degree of flexibility permitted the decision-making agencies; and (4) recommends various legislative and administrative actions to improve the rationality of the process and the substantive accuracy and credibility of regulatory decisions involving chemicals.

INTRODUCTION

Because this paper presumes the existence of a scheme of federal regulation, we will look first at how government has arrived at our current regulatory approach to the problems posed by chemicals in the environment, and at whether there are useful alternatives within our system of government. In our common law system the traditional remedy for a civil wrong is a tort action for money damages and/or injunctive relief. Individual or corporate actions that abridge personal or property rights have been addressed in the judicial forum for hundreds of years. However, drawing upon the basic power to provide for the public safety, the government has for generations required the makers of certain products, such as food, to meet a threshold governmental standard of safety and wholesomeness because after-the-fact relief by individual legal action was found inadequate. Such relief did not protect the societal interest in maintaining public confidence in the food supply, and it permitted the individual to sample the product only at his hazard. In addition, the transactional costs of consumer protection individually undertaken and the formidable legal barriers, such as discovery of the wrongdoer, lack of privity, and the sheer time and expense of the legal process argued persuasively for governmental intervention.

Similarly, the first federal regulation of chemicals not associated with food or drugs—which occurred in the area of pesticides—undertook federal review of labeling claims of both efficacy and hazard. Unfortunately, not everyone at risk reads labels, nor are all potential problems remediable by written warning. The broader federal regulation of pesticides, as well as other chemicals, has evolved through recognition that there are now many chemical substances which may affect man deleteriously even though he is not a voluntary consumer. In certain circumstances he can neither protect himself from exposure nor, in most cases, trace any physical harm which may result to a particular source. The latency of the hazard may make long-term, low-level exposure dangerous but unquantifiable and undetectable by the individual. Finally, a number of the unintended damages done by chemicals in the environment have an impact actually or potentially on animal or plant life whose protection requires a public voice.

The federal government has been chosen as the nexus for the bulk of this regulation because (a) state or local control is often ineffective or contradictory with regard to most products in interstate commerce, and (b) the media through which the chemicals move obviously do not respect political divisions. International regulation is as yet nascent because of the lack of international decision processes, the diversity of risks and benefits of particular uses, and differing scientific conclusions as to the hazards posed.

Given the perceived need for a federal role in regulation of chemicals, a system has evolved whereby the legislative branch recognizes a particular problem by statute, sets forth various standards and prescriptions, and delegates a degree of decision-making authority (and sometimes quasi-legislative power) to an executive agency. At the behest of an affected party, the particular executive decision is subject to judicial review.

While this complicated pattern is far from faultless, as discussed below, it does appear preferable to alternatives that would give one branch of the federal government virtually all the regulatory power and responsibility. For example, one conceivable division of governmental responsibilities might assign all duties to the legislative branch which then would be required to produce a thoroughgoing code covering all desired regulatory standards, deadlines, and other judgments. The executive and judicial functions would simply be to prosecute and punish transgressors. While such a system might have the virtue of clear public accountability, as well as predictability, it would totally lack flexibility and any means of translating general policy into specific directions regarding an individual substance, plant, or industry.

The challenge to government is thus to draw upon the strengths of each constituent branch—policy synthesis and political accountability by a legislature of diverse interests, adaptability and expertise of an ongoing executive agency, and objective oversight of process and implementation by the judiciary—to achieve the societal goal of protecting the citizen and the environment from unreasonable risk while minimizing intrusions upon individual freedom and ensuring the benefits derived from proper use of chemicals.

GOVERNMENTAL GOALS AS EXPRESSED IN LEGISLATION

The environmental laws are not a perfect mosaic infused throughout with harmony and purposefulness. They were developed at different times, by different committees of Congress, and they reflect all the vagaries of competing pressures and regulatory schemes. Thus, it should be no surprise that not only are related acts inconsistent, but also that inconsistencies and ambiguities exist between different sections of the same act. (The futility of imposing more rationality on the laws than they inherently contain is illustrated by the recent Court of Appeals debate over lead in gasoline [*Ethyl Corp.* v. EPA]. The majority and the dissenting opinions compared and analyzed in detail several key provisions of the Clean Air Act to determine the proper judicial standard of review for the section in dispute. The members of the court reached diametrically opposed conclusions in ranking which sections of the Act required the strictest and which the loosest review standards. Perhaps the court had no alternative to this inquiry, but its dilemma does demonstrate that the provisions of the statutes are not transparently obvious.) However, the fact that a particular chemical may sometimes be handled differently under different statutes is not necessarily illogical, because different circumstances may result in quite different risks and benefits. A chemical such as arsenic, for example, may be treated one way when used as a pesticide, another way when emitted into the air or water as an industrial by-product, and yet a third way when used as an additive or drug.

COMMON AND DIVERGENT PATTERNS IN THE CHEMICAL REGULATORY PROCEDURES

Uniformly Shared Patterns and Goals

There are a number of common theses in the environmental legislation currently governing chemical regulation. First, and probably most obvious, is the strong concern for human health. Second, national standards have become paramount. The primary responsibility for environmental regulation has shifted from local standards set and enforced by the states to national standards established by the federal government. This is because the states frequently lack the resources, expertise, and ability to withstand economically powerful special interests groups. Although the states have been given certain important roles, such as the implementation of the ambient air standards, even here they are generally following federally set standards under federal supervision.

Finally, many of the laws regulating chemicals make an implicit assumption of scientific progress, regarding as inevitable both the increase over time in scientific knowledge about the unintended effects of chemicals and the development of technology with which to control them. Additional knowledge is generally thought to be ensured by the legislative provision for government-supported research and development written into most of the statutes.

Recurrent Themes in Some Regulatory Legislation

There are some common features present in most environmental statutes but absent from a few. The seemingly obvious goal of protecting the environment is an example.[1] Another often recurring but not uniform element is the requirement for inquiry into the availability of safe substitutes or alternatives to a hazardous product prior to a regulatory decision. The pesticide act (Federal Environmental Pest Control Act [FEPCA]) allows such an inquiry, but the Delaney Amendment to the Food, Drug, and Cosmetic Act does not.

There is also considerable variation in the applicability of risk–benefit calculus to decision making under the various statutes. This is attributable perhaps less to an oversight by Congress than to a feeling

[1]See, for example, the Drinking Water Act and the residue-tolerance portion of the Food, Drug, and Cosmetic Act (FDCA), which in their legislative mandate are concerned only with the protection of human health.

that in particularly hazardous circumstances, such as the presence of carcinogens in foods, Congress itself will do the benefit–risk analysis in drafting legislation. Thus, there is no explicit statutory requirement for a consideration of economic and societal effects in the hazardous pollutant section of the Clean Air Act, and it can be argued that the legislative history of the provision indicates a specific congressional intent to exclude it.[2]

Issues Unaddressed by the Regulatory Legislation

Finally, some relevant issues remain unaddressed in the current laws. For example, the jurisdictional overlap between EPA and FDA on pesticide residues has been partially resolved by memoranda of understanding between the two agencies. Nevertheless, the relationship between EPA and the Consumer Product Safety Commission on many issues involving chemical regulation remains clouded. In 1973 a dispute arose between EPA and the Occupational Safety and Health Administration (OSHA) over responsibility for the protection of farm workers from occupational exposure to acutely toxic pesticides. It was settled in EPA's favor only after the intervention of the judiciary, individual members of Congress, and the White House.

Another unaddressed issue is the extent to which mere cautionary labels on hazardous chemicals should be sufficient to permit their continued distribution. Despite the 1972 revision of the pesticide act, there remains the legacy of earlier legislation in which merely requiring adequate notice on labels was determined to be an adequate public safeguard.[3]

DIFFERENT APPROACHES TO THE REGULATION OF CHEMICALS

As an analytical matter, the regulation of potentially hazardous chemicals may be accomplished in one or both of two basic ways: the

[2]EPA has nevertheless felt bound to justify its decisions by evaluating economic and societal effects, along with environmental effects, to demonstrate that particular decisions do not cause society more harm than good.

[3]See, for example, the Stearns' Paste case, *Stearns Electric Paste Co.* v. EPA, 461 F.2d 293 (7th Cir. 1972), which decided that a phosphorus-based home-use rodenticide responsible for scores of deaths and injuries (mostly to children) could continue to be marketed provided that it bore the word "poison" and a skull and crossbones emblem. See also the recent Alamogordo mercury fungicide case, *First National Bank of Albuquerque* v. *U.S.*—F. Supp.—(D.C.N.M. Jan. 29, 1975), in which mercury poisoning caused by eating pork fed grain intentionally treated with mercury was held not to give rise to government liability.

chemicals themselves may be directly controlled (e.g., by standards or use restrictions); or the industrial sources of pollutants may be regulated to limit or to ban discharge.

Direct Control of Chemicals

Direct regulation of chemicals may be based on at least four different assumptions. First, there can be a determination that a chemical is so dangerous that its use will not be permitted under any circumstances. This decision, which may or may not involve consideration of benefits, is best illustrated by the Delaney Amendment on carcinogenic food additives and by a final suspension decision with regard to pesticides, but it has its counterparts in the emergency sections of the Clean Air Act and the Federal Water Pollution Control Act (FWPCA).[4] However, these powers are seldom used, and, then, only after an extended administrative process.

A second, more common determination is that a chemical, although hazardous, is safe below a certain concentration.[5] This determination may be based on scientific fact or practical necessity. Where a chemical occurs naturally, where its economic utility is great, or where it is impossible to remove it from the environment over the short run, a conscious decision to use a no-effect level may be made despite lack of scientific justification. Although too often these standards are naively viewed as reflecting the scientific establishment of an actual no-effect level for a given chemical, they usually merely reflect the commonsense judgment that low dosages of a hazardous chemical are less dangerous than large dosages.

Third, there may be a finding that a chemical, whatever its toxicity, may be relatively safe if its use is permitted only by specially trained personnel. This determination is the basis for EPA's new authority to impose and enforce use restrictions on pesticides, including the requirement that certain toxic and restricted pesticides should be handled only by trained, certified applicators. Such an approach, which is more useful for acutely toxic chemicals than for those with low-level chronic effects, is often more concerned with protection of the user than of the environment.

Fourth, there may be no government regulation at all, based either on

[4]CAA section 112, FWPCA section 504.

[5]The possibility of such a determination underlies the setting of numerical standards such as the primary and secondary air quality standards under the Clean Air Act. It is also the basis for the tolerance and residue procedure under the pesticide and FDC acts.

the affirmative finding that a chemical is, in FDCA terminology, "generally recognized as safe," or—at least theoretically—on ideological grounds that business judgment or the product liability laws provide sufficient deterrents to the indiscriminate distribution of the particular hazardous substance.

Regulation of Chemical Sources

Chemicals may also be regulated at their industrial source. The chemicals in this category are frequently the leakage or by-products of an industrial process, such as manufacturing or power generation, in contrast to intentional production and use. This mode of regulation is much more common in the air and water laws.[6]

A frequent effect of regulation by point source is to stimulate the development and adoption of new technological methods of control. A sharp analytical distinction should be made between those statutes requiring the installation of adequate existing technology, and those demanding the development of completely new technology.[7]

Technology can be forced either by setting strict standards, thereby leaving the attainment to industry, or by directly specifying (when known) the technology that must be adopted. Industry has almost always preferred that the government set the goals and leave the methods of attainment to industry. In at least one case, however, EPA and the asbestos industry agreed on the procedures and equipment that should be used, but disagreed on the result that this technology could achieve. EPA should have the legal flexibility under such circumstances to require specific equipment and performance standards, rather than having to set arbitrary numerical standards. Regulation could also be based on

[6]An illustration is the Clean Air Act's new source performance standards, which prescribe emission limits for pollutants from a particular source regardless of the concentration of that chemical in the surrounding air. This indicates concern less with the direct immediate effects of, for example, the discharge of sulfates on health or the public welfare, than with the overall burden of pollutants on the environment which may result in such undesirable phenomena as acid rain hundreds of miles away, and a policy forcing uniform technology to avoid economic incentives to new development in relatively unpolluted areas.

[7]The best example of the latter approach is the auto emissions section of the Clean Air Act, which sets numerical standards so low that the industry's ability to achieve them within the statutory deadline was seriously disputed and which was subsequently extended. The Federal Water Pollution Control Act, while implicitly requiring drastic new technology to achieve its "goals" of zero discharge, seems more concerned with ensuring the installation of high-level, but existing, equipment. It mandates the use of "best available" technology (meaning the adoption of the most advanced technique then existing) by 1983.

requiring "good housekeeping" procedures in plant operation to ensure that all necessary steps to reduce pollution are taken during the industrial process.[8]

Sources of pollutants may also be regulated on a source-by-source basis, as is done in the negotiated water pollution permit program. This procedure ignores most of the ambient standard assumptions, such as reliance on a no-effect level, and concentrates instead on reducing effluent to the maximum extent technically (and/or economically) feasible at a given time.

BURDEN OF PROOF

Because the scientific evidence regarding health and environmental effects is so difficult to obtain with precision, and the costs of data collection can be so high, the party carrying the legal burden of proof is at a considerable disadvantage. This issue is quite complex, for the burden of proof can shift during different stages of the administrative process and can require different degrees of proof.

The clearest statutory burden is in the pesticide area, where agricultural chemical companies must demonstrate affirmatively that their product is safe. There can also be a negative burden, as under the hazardous pollutant section of the Clean Air Act where a substance initially listed as a hazard is finally determined to be such unless the Administrator finds within 180 days that it should be deemed safe.

The burden of proof concept is a necessary one because, first, one of the parties usually has access to most of the necessary information on the chemical and thus should bear the burden and cost of producing the evidence regarding it; second, in the many situations where the scientific evidence is tenuous or evenly balanced a procedure is essential for resolving close disputes in favor of the public interest as established by the considered political judgment of the Congress.

ADMINISTRATIVE PRACTICES

Regulatory decisions involving chemicals are the product of many factors, some of them explicitly mandated and others implicit or even unconscious.

[8]This procedure has been accepted in practice for control of some forms of mercury. A numerical emission standard was set, but because precise measurement is not possible, EPA has allowed companies to comply by following specific practices designed to result in effluent below the standard.

Rationale—Explicit Determinants

The Authorizing Statutes The first order of explict mandate comes from the authorizations or requirements contained in statutes. Further, the procedural framework within which regulation takes place, even if unspecified in particular legislation, must comply with the minimum standards of fairness set forth in the Administrative Procedures Act. These substantive and procedural requirements may, in turn, affect the less visible but nonetheless important internal agency activities affecting regulation, such as program planning and allocation of resources available to implement the legislation.[9] Another internal agency judgment affected by explicit legislative determinants is the requirement to acquire through research and development the knowledge that will enable appropriate regulatory actions to be instituted within the specified time.

Other External Stimuli Court decisions on legal challenges to particular regulatory decisions brought by industry or citizen groups or both often require the Administrator to take certain actions that may include additional research on the toxicity of a substance, monitoring or issuance of standards within a specified time, or other judicially prescribed procedures. Congress, in oversight hearings, or the President by Executive Order, may provide supplemental directives to an agency regarding implementation of existing legislation. In these instances, as well, the rationale for the consequent regulatory decision is largely explicit.

Rationale—Implicit Determinants

In interpreting or implementing legislative acts, an administrator or other decision maker is often affected by certain implicit factors that may be of either a conscious or unconscious nature.

An administrator of a regulatory agency is properly sensitive to the general political climate in which he must operate. Political pressures come from different sources: a citizen's group seeking regulatory

[9]Premarket screening of a chemical substance, as contemplated in various drafts of the toxic substances control legislation considered in the 93rd Congress, although requiring interpretation and judgment on the part of the EPA Administrator, is an example of an explicit legislative requirement affecting regulation of chemicals, as to degree of judgment, the time authorized for carrying out legislative requirements, and a particular mandate to acquire information and expertise in planning for agency support.

controls on dumping chemicals in coastal waters; political representatives of a community concerned because of imminent closing of an industry on which the community is dependent for jobs and economic welfare; an affected industry or trade association concerned about the economic impact of a particular regulation; a Congress concerned to see its legislation implemented as intended or responding to constitutent pressure; and pressures from the President or the press.

Other implicit rationales for a regulatory decision may be the desire to correct what is viewed by the agency as inadequate or earlier misconstrued legislation,[10] or to respond consistently to regulatory actions of other agencies. The understandable but pernicious desire to buttress a weak or erroneous decision by post hoc rationalization can also infect the administrative process, and it is rightly condemned when discovered. (See the lending case of SEC v. *Chenery Corp.,* 332 U.S. 194 (1947), and its progeny.)

A decision maker in a regulatory agency may be strongly affected, consciously or unconsciously, by a variety of subjective beliefs and prejudices. For example, personal political philosophies, the interagency and intra-agency interplay of personalities, or the nature of an individual's educational or professional background have a nebulous but real effect on the nature of the decisions.

Previous success or failure to obtain administrative, judicial, or congressional approval on action taken on a similar matter may also exert an effect. One hesitates to follow a once unsuccessful path without stronger motivation or some reason to anticipate success; threat to personal prestige and loss of time work against repetitions that may prove futile.

SETTING AGENCY PRIORITIES

Apart from the institutional and personal considerations affecting particular decisions, there are also a variety of factors influencing the way in which the agency chooses what to do from the broad range of its general mandates. Priorities may evolve from preliminary research results, recognized problems in other countries, or predictions from theoretical considerations. The feasibility of regulatory action in resolving potential or actual adverse effects of a chemical is one factor in assigning priority. Priorities may involve research, testing, monitoring, or

[10]An example of this is the early interpretation of the pesticides legislation which permitted only registrants to testify at hearings. This act was later reinterpreted to permit concerned groups to appear at, or even institute, administrative proceedings involving pesticide issues.

staff time required to achieve some level of regulatory control. Other considerations may include some level of regulatory control as well as explicit judgments about the commitment of the agency's resources, benefits accrued, costs to society, and other factors. The most readily attained result in terms of cost and time will affect the decision maker in assigning priorities to a chemical that has been identified as a potential hazard for one or more of its intended uses.

Outside pressures often affect assignment of priorities. These may involve pressures from scientific advisory groups, citizens groups, industry, or Congress. Other pressures may result from internal or external research findings that show chemicals to have potentially adverse environmental effects. Pressures from other executive agencies may also affect a decision to restrict a chemical.

REGULATORY OPTIONS

In the decision process, the responsible administrative decision maker has at least five options to consider in the regulatory control of a toxic substance. These include (1) the decision to regulate; (2) the failure to make a decision (conscious or unconscious); (3) the affirmative decision not to regulate; (4) the decision to deregulate; and (5) the decision to permit limited or voluntary regulation (negotiated decision).

In the first category, the decision to regulate is generally based upon a benefit–cost analysis with risk clearly the dominating factor.

In the second category, the conscious decision not to regulate a substance known to be hazardous or potentially hazardous may result from ignorance, or partisan political or benefit–cost reasons, or a combination of these factors. An example of nondecision that may be unconscious rather than conscious is the lack of regulation for cadmium in spite of the growing evidence that this metal and its compounds are a hazard to humans and certain other organisms.

The third category, an affirmative decision not to regulate, applies to the bulk of new chemicals. Nevertheless, the regulatory agencies must commit a substantial portion of their resources to benefit–cost evaluation of these substances to assure the concerned public that the decision not to regulate is a safe and proper one.

The role of the decision maker in reviewing a regulated chemical for partial or full deregulation, the fourth category, should be basic to the function, philosophy, and objective of the mission of the regulating agency. Any decision to regulate a chemical should remain in effect only as long as the reasons for having controls are more compelling than those for reducing or eliminating them. Generally, the initiative and burden of proof in the deregulatory process lies primarily with the manufacturer or user. Once a substance is regulated, however, it is rare that it is completely deregulated.

Voluntary regulation or self-regulation, category five, is often desirable from the viewpoint of both the regulatory agency and the manufacturer whose substance is regulated. If it can be demonstrated that a substance is hazardous or potentially hazardous, the manufacturer may elect to withdraw this substance from the market for those uses that are or may be hazardous. The voluntary withdrawal of PCBs from uses in the food processing industry, NTA from detergents, lead from interior house paint, and the reduction of antifungal mercurials in paint are examples of voluntary controls negotiated with regulatory agencies.

THE DEGREE OF AGENCY FLEXIBILITY AND DISCRETION

THE EXISTING STATUTES

Currently there exists wide variation in the discretion delegated to administrative agencies by legislation regulating toxic chemicals. The spectrum stretches from the Delaney Amendment to the Federal Food, Drug, and Cosmetic Act [21 U.S.C. sec. 348(c)(3)(A)], which flatly prohibits food additives shown to be carcinogenic in appropriate animal experiments, to the wide discretion granted EPA under the Clean Air Act to regulate fuel additives (see section 211, whose legislative history, however, reveals that the drafters arguably intended to regulate lead, at least; the section itself does place some limits on the Administrator's discretion) and to promulgate and set time limits for achievement of secondary ambient air quality standards (see section 109).

There are several variables affecting regulatory decision making which may be left to agency discretion. Certain statutes delegate to the relevant agency some or all of the substance of the standard itself, and/or the criteria by which that standard is set (see section 6 of FEPCA, for example). Timing also constitutes an important variable which may or may not be delegated, both as to specific time periods within which standards are to be promulgated and the sequence of events which is to create and/or enforce them (see section 301 of FWPCA, section 112 of the Clean Air Act, and section 6 of FEPCA). Some statutes delegate to the agency the decision about what type of procedure—adjudicative or quasi-legislative—is to be used in producing regulatory standards; other statutes do not (compare section 6 of FEPCA with section 307 of FWPCA). Certain statutes (e.g., the Solid Waste Disposal Act as amended) merely grant powers, while others (e.g., section 306 of FWPCA) contain affirmative mandates to act.

DETERMINING REGULATORY AGENCY DISCRETION

There are a number of considerations about decision making by regulatory agencies that should affect a decision as to what constitutes an appropriate degree of delegation of authority to agency discretion. One is the need for predictability. Specific legislative standards are predictable, at least to their authors, while standards to be derived from delegated authority are not. (Predictability is particularly important insofar as it affects private sector economic and technological planning requiring massive investment and long lead times. Automotive emission standards are the most obvious example.) Another consideration is the need for prompt decisions: administrative agencies may be slow to act without a statutory or other external stimulus. (Pesticide cancellation proceedings before EPA and various food additive and vitamin hearings before FDA, for example, have been notoriously protracted over many years.)

Specific accountability is present to a much greater degree in administrative than in legislative decision making. Yet the extent to which the administrative decision maker necessarily relies for advice upon "invisible" subordinates gives opponents of the ultimate decision an opening to challenge the objectivity and wisdom of the alternate determination.[11] However, the legal requirement of adequate articulation and explanation of the basis for an administrative decision and the final accountability of the decision maker eliminate the potential harm of this inevitable administrative practice. The threat of invoking the Freedom of Information Act to obtain documentation as to how a decision actually was made is an additional check upon anonymous and unarticulated agency regulatory decisions (*Montrose Chemical Co.* v. *Train,* 491 F.2d 63 [D.C. Cir 1974]). Other types of regulation constitute sufficiently important national policy decisions to be most appropriate for resolution in the first instance by elected representatives (see, for example, section 301 of FWPCA and section 202 of the Clean Air Act, dealing with effluent and emission control, respectively).

Public acceptance of any standard is essential. Tne credibility and legitimacy of a standard depend upon both its source and the procedure

[11]EPA's recent DDT cancellation and Aldrin/Dieldrin suspension decisions were challenged in the courts on this ground. See, for example, EDF v. EPA, 489 F.2d 1247 (D.C. Cir. 1973).

used in its derivation. The more far- reaching the regulation, the more appropriate it becomes for legislative determination, at least regarding policy direction. The more uncertain the scientific basis for regulation and the greater the need for flexibility and adaptability, the more appropriate administrative determinations become.

The type of decision-making process envisaged may affect the degree of agency discretion considered appropriate. Agencies and their administrative law judges are better equipped to conduct formal adjudication than is Congress. The more scientifically complex the regulatory decision, the more agencies are the preferred decision makers, even in quasi-legislative rule making, since their technical expertise is greater and support staff more extensive. Further, their actions are ultimately subject to judicial review. This vital check against arbitrary and capricious regulation—largely absent as a threat to legislative regulation short of Constitutional challenge—is an important factor which should encourage greater delegation of complex, continuing, and highly technical regulation to administrative agencies.

Another factor regarding delegation of standard-setting discretion to regulatory agencies is the relevant agency's sensitivity to partisan political concerns. Different agencies have differing degrees of insulation from the political process; at least in theory, the "independent" regulatory agencies have greater immunity from partisan politics than do executive agencies directed by Presidential appointees. Arguably, the broader the regulatory policy decision and its effects, the more appropriate that it reflect the will of elected officials or those appointed by them, hence ultimately the voters.

Other characteristics of regulatory agencies are important in considering how much discretion should be delegated to them in standard setting. Agencies frequently cater to their own institutional self-interest, budgetary and otherwise, rather than the public interest. They frequently develop over time too close a relationship with the economic interests they are supposed to regulate.[12] Bureaucratization, with its attendant evils of procedural complexity and inertia and the enormous expenditure of time and money by those involved in the agency standard-setting process, appears to be as inevitable as it is undesirable.

[12]Examples include USDA's reluctance to regulate pesticides while at the same time promoting them—a situation which contributed to EPA's creation and Presidential transfer of pesticide regulation and enforcement to that agency, and the decision to split the AEC, with its joint promotional and regulatory authority for atomic power, into two separate agencies (Energy Research and Development Agency and the Nuclear Regulatory Commission).

Finally, the level of government involvement is important. Federal regulatory agencies have more adequate funding and staff than is usually available at the state and local level where delegated regulatory authority may be unexercised merely for want of resources.

RECOMMENDATIONS FOR CONSIDERATION BY THE COMMITTEE: NEW DIRECTIONS—TOWARD A BETTER LEGAL FRAMEWORK

The objectives of government in the regulation of chemicals should be to protect public health and the environment, to minimize the costs and complexity of regulation, to ensure the credibility of regulation by making the decision process rational and open to the public, and to permit enjoyment of the beneficial uses of chemicals. After reviewing existing statutes and patterns of administrative action in this area, the panel concluded that Congress in enacting chemical legislation and the agencies responsible for implementing it have fallen short of these goals in the areas discussed below.

RECOMMENDATIONS TO CONGRESS

Congress should set more consistent and specific standards to guide the agency in its regulatory decisions and the courts in their review of agency determinations. More explicit direction is critically needed in four areas:

Benefit–Risk Analysis

Congress has set few coherent guidelines on the extent to which the benefits, as well as the risks, must be considered in government regulation of chemicals. In some cases, Congress, viewing the issue of socially acceptable level of risk, has concluded that it is essentially a policy question involving basic social values and, undertaking the balance itself, has been very explicit.[13] In other legislation, Congress has explicitly given the agencies broad discretion to make their own benefit–risk judgments.[14] In still other acts, the scope and even appropriateness

[13]In the Delaney Amendment to the Food, Drug, and Cosmetic Act, for example [21 U.S.C. sec. 348(c)(3)(A)], Congress made the decision that no conceivable social or economic benefit can outweigh the risks from human ingestion of carcinogenic food additives.

[14]In the FEPCA, for example, pesticide manufacturers may, upon showing of substantial economic benefits, demonstrate that the agency should alter its initial finding that a pesticide poses an "unreasonable risk" to human health.

of agency benefit–risk analysis are left unclear, or the concept is simply not addressed.

It is this panel's judgment that benefit–risk analysis—defined simply as full and unfettered consideration of all material relevant to the regulatory decision—is a useful tool in helping to make chemical regulation more rational and efficient by explicitly permitting full consideration of all economic, social, and environmental information. Congress, in mandating this analysis, should determine how objective and publicly credible data on risks, control costs, and economic benefits may best be obtained from government and industry.[15]

Burden of Proof

Congress in its statutes and the agencies in their rules of practice should clarify the question of who should have the burden of proof at every stage of adjudicatory proceedings. In regulation of chemicals, the necessary scientific evidence is sometimes equally balanced or unknown. Under these circumstances, assignment of the burden of proof will ultimately determine the outcome of the contested decision. The panel believes that Congress should assign the burden of proof so that: (1) it will be clearly understood in advance of regulatory action and thereby permit disputes to be settled more equitably, expeditiously, and predictably; and (2) it will have been set by the most politically accountable decision makers.

In certain statutes (e.g., FEPCA) the burden of proof is explicitly on the private registrant to establish the safety of his product, but this specificity appears to be an exception to the general rule. In most cases the burden of proof is, at least by implication or judicial gloss, on the opponent of continued use of a chemical.[16] Until this burden is met, release of the chemical into the marketplace (and the environment) continues. The public may therefore be exposed to the risk of a

[15]The panel recognizes that benefit–risk analysis is no Rosetta Stone for deciphering the mysteries of chemical regulation. Such analysis is critically dependent on the availability of an independently verifiable data base. Industry has an incentive to exaggerate the control costs of regulation and the economic benefits of chemical products. Overzealous citizen groups, congressional committees, and agencies may exaggerate the risks. Compounding these problems is the analytical and philosophical difficulty of weighing concrete economic costs against often unquantifiable long-term hazards and the assignment of a quantitative value to a human life, environmental concern, or aesthetic deprivation.

[16]See, for example, the Federal Water Pollution Control Act, as interpreted in the *Reserve Mining* case, 498 F.2d 1073 (8th Cir. 1974) *aff'd in part and rev'd in part*—F.2d—(8th Cir. March 14, 1975).

reasonably suspect chemical for an indefinite period of time while regulatory decisions are made.

It appears more equitable for the vendors of a chemical, who profit from its use and have superior access to pertinent data and other research resources, to carry a heavier evidentiary burden in these proceedings than the public (or the government acting for the public). Further, the vendor's costs in studying the product are passed on to users of the chemicals through market processes, thereby spreading the costs equitably among the beneficiaries.

Therefore, this panel recommends that in adjudicatory proceedings Congress direct that once a *prima facie* case has been made that the challenged use creates an unreasonable risk to human health or the environment, the burden of producing evidence should shift to the proponent of use, who must then make an appropriate showing that the chemical is safe on balance. (Congress must direct the agencies and the courts as to the appropriate standard to apply, such as "substantial evidence," "preponderance of the evidence," or "beyond a reasonable doubt.") In determining the standard of evidence which an opponent of use must satisfy to make a *prima facie* case, the immediacy or latency of the hazard and the benefits as well as the risks of the chemical are factors for which Congress must provide guidance for agencies in the particular case. When this standard has been met and the burden shifts to the proponent of use, Congress should provide for an expedited, time-limited administrative hearing.

This process for allocating the burden of producing evidence should apply, in the first instance, to the premarket stage of decision making prior to widespread use of a chemical (see FIFRA, FDCA, and the proposed Toxic Substances Control Act), as well as to enforcement proceedings against chemicals already on the market.

Judicial Review

Congress should clarify the intended standard of judicial review of chemical regulatory decisions. Certain statutes prescribe a "substantial evidence" test (OSHA, FIFRA), but most statutes are silent on what the appropriate standard should be. Yet this crucial determination can decide the issue. To set specific standards, Congress must make a value judgment as to the proper role of administrative discretion. In quasi-legislative administrative proceedings, such as rulemaking, a "preponderance of the evidence" test may be required. In adjudicatory decisions, which have built in the more comprehensive procedural safeguards of the adversary system, a substantial evidence or "arbitrary and capricious" standard is often applied. In any case, to increase the predictability and consistency of judicial review, Congress should articulate what standard of judicial review it intends.

Delays

Congress should review those adversary proceedings which have become unduly protracted, such as the cancellation hearings under the FEPCA and the years of litigation over Reserve Mining, and determine how best such marathon decision making can be compressed. A deadline (which the Administrator could extend for good cause) might be specified, or some penalty might be established as an incentive for any party not to delay unreasonably.

The panel believes there is an urgent need for congressional enactment of a comprehensive Toxic Substances Control Act to address many of the problems and recommendations set out herein.

RECOMMENDATIONS TO EPA

1. There is a need for EPA to make its decision making more planned and less reactive to outside events. The panel makes two suggestions in this area:

a. Greater use should be made of committees of experts, appointed to ensure that their makeup represents the spectrum of potential viewpoints, primarily for long-range planning and in helping to anticipate future chemical control problems.

b. A matrix or generic approach, as opposed to an ad hoc procedure, should be adopted whenever scientifically possible for the regulation of chemicals. (For example, when a hazard from a particular pesticide is determined, EPA should attempt, to the extent its resources permit, to examine the other members of that pesticide class at the same time. To investigate the hazards of 2,4,5-T, for example, without examining other phenoxy herbicides is both inefficient and needlessly risky.)

2. Regulatory decisions affecting chemicals require either more insulation from political pressures narrowly defined or, alternatively, more complete and systematic exposure to the political process broadly construed. The former approach would enable EPA expertise to be brought to bear on regulatory decisions without overt pressure from the White House, Congress, or special-interest groups; the latter approach would expose these pressures to public scrutiny and would guard against regulations by anonymous or parochial decision makers, unresponsive and unaccountable to society.

Unlike the independent regulatory commissions, EPA is an executive agency with an Administrator who serves at the pleasure of the President. This structure of a unitary decision maker at the head of the Agency should result in the maximum public accountability possible in nonelected officials. This panel believes that a single Administrator is strongly preferable to the collegial decision making of the independent regulatory agencies.

To increase further the objectivity and openness of Agency decision making, the panel makes the following recommendations:

a. Any adjudicatory decision pending before the Agency should be made public with sufficient time for comment before a decision is made. (This is now required in EPA's regulations implementing the FEPCA. 40 CFR sec. 164.7.)

b. A more comprehensive description of the technical data and expert advice relied upon by the Administrator in making his decisions should be made public, wherever this is not currently done. (In appropriate circumstances, as when the Agency has asked for public comment on quasi-legislative decision making, relevant information available to the Agency should be made publicly available sufficiently in advance of a final decision to permit considered response. On a more continuing basis, thought should be given to establishing a process, source, or institutional center whereby comprehensive information on toxic substances would be readily available to the public.)

c. The basis for "nondecisions," including the conclusion to exempt a chemical from regulation, should be made public when a decision is made not to go forward with a regulatory effort.[17] Agency regulations should define those circumstances where this is advisable.

d. To ensure the objectivity and public credibility of the scientific evidence presented in adjudicatory proceedings, the right of counsel to cross-examine expert witnesses (such as scientists and economists) must be guaranteed. FEPCA now grants EPA subpoena power to call expert witnesses. This power should be extended to proceedings under other statutes in order to take advantage of the adversary process and to develop the fullest and fairest public record. This goal must, of course, be tempered by the practical necessity of making decisions without endless procedural delays.

[17]In EDF v. *Hardin,* 428 F.2d 1093 (D.C. Cir. 1970), the courts took a major step in this direction by requiring agencies to state the basis for nondecisions after an appropriate interval or be judged to have acted in an arbitrary and capricious manner. See also CEQ regulations reflecting judicial interpretations requiring an agency to explain a decision not to file an environmental impact statement under NEPA. 40 CFR Part 1500.

APPENDIX D

Working Paper on the Role of Nonfederal Participants in the Regulatory Process: The Public or Publics

SUMMARY AND RECOMMENDATIONS

The U.S. Environmental Protection Agency (EPA) was created in response to the environmental movement of the 1960s led by environmental organizations, certain segments of the scientific community, the news media, and particular groups within organized labor and industry. Being new, EPA had no established technical or administrative relationships with other federal agencies and had to develop from within, while limited by statutory, executive, and, before long, judicially established boundaries.

The Agency is young and still developing. During its first years, it has had a succession of changes in top-level staff, hierarchical organization, and methods for reaching decisions. Because of these changes, predictable actions and consistent decisions have not been an EPA hallmark.

This lack of structure and consistency in decision making was also a factor in EPA's reluctance to open its operations to the news media and public scrutiny. But it is now time for the Agency to open its decision-making process to interested parties whenever feasible. Many knowledgeable persons and the nonfederal publics they represent are willing to share their expertise and informed judgment with EPA in a more responsible, responsive, and flexible manner than has been possible until now.

How then can the government agency responsible for environmental decisions assure that these several regulated publics, and the general public in whose interest the regulation is undertaken, have optimum opportunity to participate in the decision-making process?

RECOMMENDATIONS FOR CONSIDERATION BY THE COMMITTEE

1. The decision-making process must be structured or formalized in ways that assure increased openness and receptivity to the knowledge, expertise, and considered judgments of the nonfederal publics directly involved. While the process must become more formalized, it should ease and not restrict the accessibility of every responsible and concerned citizen to the entire apparatus and structure for decision making. Formalization means, at the minimum, a process described and published in the *Federal Register* for various steps in the decision-making process.

2. Before new regulations take effect, responsible officials should consult with a broad variety of nonfederal publics at two preliminary stages in the decision-making process: first, when a regulatory action is first considered, and second, after the marshaling of the pertinent facts and before the decision becomes final. Active solicitation of suggestions and comments is essential, and mere publication in the *Federal Register* is inadequate. Informed participation of the news media in eliciting public cooperation and comment is highly desirable. Special lists of nonfederal publics with particular interests should be used to invite responses.

3. Continuing education of the general public and the more directly involved publics on problems of chemicals in the environment and upcoming decisions by the EPA is important to a broader and deeper understanding of the pertinent laws and regulations and the decision-making process itself. An affirmative education program by specially trained and selected outreach officials should be established.

4. Education of decision makers is an equally important continuing requirement for good decisions. Organized labor, businessmen, professional societies, and public interest groups must share their information, data, experience, and technical expertise with government officials, their own constituencies, and the general public. Otherwise, the deleterious economic impacts of decisions about environmental chemicals may fall disproportionately on small corporations.

5. Consideration of an independent Commission on Toxic Substances is warranted. A commission of qualified experts would make decisions on certain specified standards, tolerances, informational

requirements, and prohibitions. It would not be involved in research or regulatory enforcement proceedings.

INTRODUCTION

With the proliferation of population and technology, mankind has the capacity to destroy itself directly with military weapons or indirectly through disruption of natural ecosystems. In distinguishing societal from human evolution, it is clear that intelligence and cooperative human effort can be and have been brought to bear on technological and social problems which threaten either the quality or length of man's life.

Society's first attempts in this direction were in support of the biological goal of human survival: assurance of food supply; elimination of disease; and protection from predators, natural elements, or hostile human societies. The individual, social, and governmental processes for dealing with these problems have been the object of much effort throughout the history of civilization.

War, famine, and disease still occur. Individuals and societies still disagree about how best to cope with these problems, but there is far more agreement than disagreement. However complex the social issue, the average citizen feels competent to express his choice and act on it as an individual, as he does each time he votes.

The average citizen cannot, however, readily cope with the widely divergent views about possible environmental controversies (eutrophication of certain waterways versus tax costs of sewage treatment plants, for example). Here, he relies on government; and government is still in the process of developing guidelines and procedures for dealing with environmental issues. These issues not only affect human health, they also involve changes in the ecosystems necessary for man's survival and changes that may involve future generations through mutation of human genes.

That the average citizen may not have detailed expertise in these areas does not mean that he does not wish to be consulted. It does not mean that he trusts scientists, career officials, or elected or appointed politicians to make these decisions for him. The average citizen in the mid-1970s has a growing distrust of all these groups as being self-serving, or at least, as serving special interest groups. Some citizens, on the other hand, are content with the current process.

The challenge then is to increase the public's knowledge of environmental issues, and in so doing to improve environmental decision making and thereby increase the public's confidence in the wisdom of government's environmental regulatory decisions and policies.

PUBLIC GROUPS

The "public" is not a homogeneous group. There are multiple publics concerned with environmental issues, and these publics may all wish or need to have appropriate representation in the decision-making process. These publics include industry, labor, the press, the scientific community, public interest and environmental groups, and local and state governments. Because these six groups have special interests, expertise, and capabilities in environmental problems, we will consider them later as a subset to *the public* and first address overall public participation in the regulatory process.

For our purposes we take the view that elected and appointed governmental officials do indeed represent the general public. Further, to the extent that there is a formal process by which the six special public groups listed above may contribute to the regulatory process, we take the view that the public interest is served.

We take as "given" that an informed, confident public is better than an uninformed, angry public. We also accept the view that there must be a major role for interested members of the public in environmental decisions and that it should be a positive and creative role. An emotional "environmental sloganeering" campaign which demands and achieves unwise environmental decisions can do as much environmental damage as a public free to prescribe drugs or perform operations on one another without medical education and medical licensing laws. We must develop an environmentally educated public and simultaneously structure the decision-making apparatus so that public participation will be useful, timely, and constructive.

Public confidence is not won easily: it requires that all groups who participate in the regulatory process behave in a responsible fashion. Primarily, the change most desired is more openness on the part of the Executive Branch of government during the process of collecting and evaluating environmental data. The public needs a specified series of time frames during which interested groups can provide input into the regulatory process. Their first opportunity would be at the outset, via an open or public hearing, and then, after all data have been reviewed, they would be able to participate at a second open or public hearing prior to the final decision.

In addition to greater openness on the part of federal governmental regulators, we have a number of suggestions for improving the environmental awareness and knowledge of the average citizen as well as suggestions regarding the desirability of insulating the decision maker from unwholesome political influence.

INDUSTRY

For convenience, it is useful to divide industry into three components: large corporations, small business, and trade associations. Current input from industry to the decision-making process is, for the most part, haphazard. The three major components vary significantly in their involvement, and within each of the three groups there are even greater differences. In addition, personal relationships, past associations, and other informal arrangements rather than the existence of any policy or structural procedure provide the most frequent bases for contacts.

Currently, large corporations and trade associations are virtually the only sources of input. Small business appears to be involved very rarely, and then only when a specific product, special manufacturing process, or unique method of distribution is at stake.

Since the three components of industry vary in their ability to provide different kinds of input, a process must be devised for enhancing the positive input from each component by overcoming procedural and other limitations in the current system. To aid in the development of such a process, the specific kinds of information that each component of industry could be expected to provide are outlined below.

Large Corporations

The following kinds of data, information, and participation related to toxic substances could be expected to be available from large corporations:

1. data on physical, chemical, and toxicological properties of chemical agents;
2. data on the reactions of chemicals and their products and, to a lesser extent, the environmental fate of materials resulting from manufacturing processes and distribution of products;
3. industrial hygiene measurements of the occupational environment and measurement of agents in areas immediately surrounding production and storage facilities;
4. employee health records for mortality and morbidity studies, and their correlation with environmental exposure experience;
5. ability to bear the economic costs of participation in the decision-making process; and
6. service as a resource for preparation of materials for laboratory testing by government, academic, or private laboratories, and as a

research capability to advance the state of the art in discrete, well-defined areas as well as in broader, conceptual areas.

Small Business

Small business is severely limited in its ability to participate in the decision-making process. Its positive input would be confined primarily to special situations in which the small corporation itself is intimately involved (for example, cancellation of a pesticide it manufactures). It is important to recognize such limits. In general, these are:

1. knowledge of the state of the art in the relevant environmental sciences ranges typically from minimal to nonexistent;
2. availability of and capability for environmental monitoring and employee health and safety surveillance is usually limited; and
3. capability to bear costs of participating in the decision-making process is limited.

In addition, we note that monitoring and surveillance of small business would require a federal bureaucracy of staggering size.

Trade Associations

Trade associations currently play a significant role in the decision-making process. They can offer the following input:

1. a pooling of resources that permits support of research, purchase of consultation services, and similar functions that are beyond the capacity of single small corporations; and
2. broadly based facilities and cooperative or collaborative efforts that maximize the coherence in such participation.

There are, however, certain problems in the current role of trade associations which are important to recognize:

1. only within the recent past have certain of the trade associations given appropriate priority to matters of health, safety, and environment;
2. they take positions that are of necessity consensus positions, usually at the level of a common denominator: this may insulate the agency from information an individual corporation is developing at the leading or cutting edge of a relevant field; and
3. they are slow moving, as a rule.

Suggestions

Despite the obvious desirability of involving all components of industry in the decision-making process, this ideal situation is unachievable under existing conditions. Currently there can be no doubt that environmental decisions requiring compliance by all corporations will find their initial maximal returns in the larger industries. Public hearings, multicategorical and broadly based advisory committees, and processes for inclusive constituency participation are techniques that still provide for the most effective involvement of industry, large and small, in the decision-making process. It is, however, incumbent on the regulators to actively and aggressively seek out and involve those small corporations which would be affected by and/or could contribute to the decision-making process.

Action Proposals for Consideration by the Committee

1. In the decision-making process, EPA should actively seek out industrial involvement at the earliest feasible time (at the level of early policy or program decisions) for scientific, technological, and economic input.

2. EPA should maintain a complete roster of industrial companies, large and small, which produce, handle, or market toxic chemicals, for the purpose of assuring maximum input, particularly in those areas where commercial enterprise is engaged in part or entirely in environmental controls. There are many companies whose activities are based on the development, manufacture, and sale of environmental control equipment, and, despite "potential" conflict of interest, EPA should avail itself more thoroughly of the expertise in these companies.

3. EPA should maintain contact with business organizations and trade associations at all levels—federal, state, and local—when appropriate for the purpose of communication and education in both directions.

4. EPA, in both its statutory advisory boards and its administratively appointed committees and councils, should assure appropriate industry representation at the various stages of the decision-making process.

5. EPA should develop an educational resource particularly aimed at small businesses to help provide them with the information and access to data at present routinely available to large corporations and major trade associations.

6. EPA should seek out agencies at the federal, state, and local level

with primary concern for the small business sector, to develop cooperative and collaborative programs to facilitate participation in the decision-making process.

7. Each company should provide EPA with a list of chemicals currently being used in the manufacturing process so that an inventory of potentially toxic substances is readily available.

LABOR

It is important to recognize that most workers in the United States are not members of unions; further, several additional millions are organized in small independent unions that do not have significant professional resources. Because of this, the primary input of the worker to EPA activities is through the political process and the myriad social and community organizations in which the worker participates. This input, by its nature, is diffuse and uncertain.

A special obligation, then, falls upon organized labor to represent the interests of all workers. In this connection, the leadership and the professional secretariat of the AFL–CIO and its departments, as well as the UAW and the Teamsters, have, especially in the last five years, established strong collaborative links to the academic community, developed training programs for local union officers, and employed an expanded spectrum of specialists in health, safety, and environment.

At the present time, labor is in a position to recommend or provide expertise at every level of the decision-making process. Although the labor establishment does not resemble the institutions of industry, academia, government, or even the environmental and voluntary health movement (in which it played a founding role and remains an active participant), its input is no less needed or valid.

There are in fact unique inputs provided by the labor movement: an evaluation of the impact of alternative regulatory options on labor, definition of high-risk populations, case-finding, mortality and morbidity data, economic impact information, control techniques, determination of industry-wide practices and processes, plant-level intelligence, and the experience of administering health maintenance organizations. Labor has increasingly developed skills in statistical analysis, macroeconomics, industrial engineering, and hygiene, as well as in environmental management, that would enable it to make even more substantial contributions to EPA's decision-making process.

Finally, the most difficult decisions to be made by government will not be scientific or technical in nature: social and political decisions will be made that will channel and shape the development of our culture. In this process, labor's contribution will be unique. Alone among American institutions, it is both most exposed to environmental insult in the community and in the shop and most vulnerable to the economic consequences of control. It is, therefore, in a position of forced objectivity and thus a critical voice to be heeded.

The participation of organized labor is not automatic, however, precisely because it is in this position in virtually every crisis in our society. A positive effort must be made by government to establish a close working relationship with the leadership and professional secretariat in the labor movement. This can be accomplished by such techniques as the establishment of full-time labor liaison personnel, representation on advisory panels, and the guarantee of equal participation with other nonfederal groups in the provision and review of data used in decision making by government. When decision-making boards or panels include nongovernment representatives, labor as well as management should be represented to assure equity and balance.

The achievement of this participation would represent a change in past policies and remove another barrier to environmental sanity.

NEWS MEDIA

A free and investigative press can serve as an important vehicle for ensuring that a decision-making process is representative of the broad diversity of public opinion. To the extent that it both affects and reflects views of various segments of the public, the press provides a two-way mirror for the legitimate aspirations and opinions of citizens—organized or unorganized, informed, uninformed, or badly informed.

By influencing public opinion, the mass media can play an important role, favorably or unfavorably, in governmental decisions on regulation of chemicals in the environment. Drawing on both external events and the internal processes of news gathering, this role can either help to accelerate the decisions at a pace faster than the regulators would prefer, or it can do little to affect the standard-setting and review procedures.

The press, including television and radio, has the capacity to inflame public opinion or to calm it. The government is in a position to help direct that capability through a policy of candor and frankness, taking the press into its confidence at various stages in the decision-making process with regular briefings, interviews, and a free flow of communication. Such a policy would make the public partners in, rather than, as they often perceive it, victims of environmental regulations.

Information on suspected environmental contaminants comes to the press in a variety of ways. Papers presented at scientific meetings and in journals represent but one source. Public interest and consumer groups, each with an awareness of the critical importance of public exposure of their views, make contacts with the press a top priority.

But more often than not, information comes to public attention through some ecological calamity (for example, a fish kill from pesticides, vinyl chloride deaths, or lead-poisoning cases) reported to official government agencies—local, state, or federal. These are often fragmented and disjointed bits of unconfirmed information. It is at this juncture that the reporter, suddenly confronted with what appears to have the ingredients of a good news story in which the public is victimized by an industrial accident or industrial greed or negligence, is motivated into action. The reporter may bring no preconceived ideas of scientific rightness or wrongness to the story, but he may nevertheless perceive himself as the representative of the public (which indeed he is) and as an ex officio environmental protector.

If he follows the normal pattern, he will attempt to check the story with some official source—most frequently the local environmental protection unit, wherever it may be located in the government. More often than not, the official may know little about the contaminant or the case and be unable to give the reporter much help. The reporter will then turn to the company or industry involved. Here he may find someone who knows a great deal about the contaminant but is unwilling to help him.

At this point in the news-gathering process, both the government and the industry have been put on the defensive and have their guards up. It should be remembered that this procedure frequently takes place within a few hours. Frustrated and often faced with the tyranny of a fast-approaching deadline, the reporter writes only as much as is reported to him and usually includes a reference to the fact that neither the government nor the company involved had any comment. (Sometimes circumstances permit a reporter to wait until the immediate crisis subsides and a more balanced presentation can be made.)

When the story is read or seen on television, a public reaction of combined anger and frustration often results. Mixed with the current epidemic of mistrust of government generally, the news tends to widen the gulf between what the public expects of government institutions and what those institutions are actually doing, according to the paper or the TV screen.

A series of environmental laws, dating back a decade or more, has created a set of environmental expectations among the public. When they read in their local newspapers that nobody is doing anything about some ecological adversity or that an official has no comment on it, their confidence erodes. If public attitudes are hardened in this way, the neat timetables established for standard-setting in current or proposed legislation could be upset; or the public could be unwilling to support the standard or its implementation if it involved serious discomfort or costly changes in life style without clear compensating benefits.

Reporters generally pride themselves on their independence and insist on the right to cover a story as they wish. But, as the public's representatives, they do not want to be uninformed or misinformed. Within that framework, therefore, there is frequent opportunity for the government, in considering the impact of chemicals on the environment, to communicate freely and frequently with reporters in both formal briefings and one-to-one coverage, what it is doing to protect the public. It should not be forgotten that in interviews with perceptive reporters the government official is often able to sense what attitudes have formed among the reporter's readers or may be produced by the proposed regulation.

Regulation of toxic substances is, by its very nature, a story that will evoke sympathetic reactions from the press. It is a story with a built-in advantage for the government, which is, after all, engaged in the business of protecting that same public the press sees itself as representing.

As a minimum guide, the press should be told when a standard-setting process begins and when the proposed standard is ready for publication. Further discussion during this process should occur as often as warranted. To the maximum extent possible, the media should be invited to witness the meetings in which regulatory actions are being considered and otherwise given full access to the data on which those decisions are based.

SCIENTIFIC COMMUNITIES

The scientific communities (natural and social) comprise five major groupings of professionally trained persons: academic research scientists and physicians, industrial research scientists, physicians and other health professionals in private practice, contract research and development organizations, and professional societies. These groups, working alone and together in a variety of ways, play a major role in providing the information needed for reaching decisions on chemicals in the environment.

Public Education

Over a period of many years, the scientific communities must teach all who can and will learn that (a) no material is safe, and every compound can be fatal if one is exposed to excessive quantities of it, and that (b), the presumed converse, there is a relatively safe intake or degree of exposure for every noncarcinogenic compound no matter how deleterious or dangerous it may appear. Once there is widespread understanding of this fundamental principle, that for every substance there is a broad continuum which ranges from "safety" to lethality, the task confronting the decision maker will be significantly eased.

This role as educator of the public is the first and most essential for the scientific communities to accept and carry out. It is not easy, and it stands outside their accustomed role as investigators and purveyors of highly specialized knowledge. But some immediate benefits could be hoped for; and the long-term (10 to 25 years) and the medium-term (3 to 10 years) effects will surely be beneficial, as the citizenry gains increased perspective and becomes less susceptible to manipulation by emotional appeals that overlook the quantitative relationships between cause and effect for a particular chemical or class of chemical compounds.

Data Preparation, Development, and Presentation

Research physicians and laboratory scientists know and best understand their roles as developers of original concepts and detailed data regarding new chemicals and their potential usefulness. New drugs, food additives, and pesticides are the three classes of compounds best known and understood by scientists and clinicians when first marketed.

But because these and many other new chemicals are usually protected by patent applications, research reports providing basic information on the chemistry, mode of action, pharmacology, toxicology, clinical effectiveness, unwanted side effects, hazards in use, and so on, for a particular substance may not precede its market introduction by very long. Often these reports appear in technical and medical journals following market introduction of the new product. Under these circumstances, the government decision maker and the commercial sponsor of the new material must rely almost entirely on the data and reports gathered by the sponsor.

A particular weakness in this process is that the peer review inherent in the editorial refereeing of research articles submitted for publication in medical and scientific periodicals has frequently not yet occurred when a decision is needed. The decision maker would surely feel more certain knowing that in significant respects his judgment coincided with that of the editors and referees of several published research reports.

Another type of situation may arise where, for good reasons, the availability of a new chemical (or the development of a new use for an existing compound) does not merit the time, effort, or space required for publication of new studies, even though the commercial sponsor may have undertaken thorough and extensive evaluations to assure himself of the pertinent qualities and usefulness of the new material or application.

If time permits, and the decision maker is uncertain of the appropriate decision based on the data before him, the scientific communities can and do make a substantial contribution by providing experts in the pertinent discipline to review and comment promptly on the data at hand. This can greatly aid in the achievement of a carefully considered decision.

Initial Information on New Chemicals For new substances being introduced for the first time into wide public use, these two roles—the development of all needed scientific, technical, and clinical data, and the evaluation of these data by independent experts—are the most important functions of the scientific communities. They serve an essential role in the immediate task of reaching a particular decision regarding a specific substance.

Although the accumulation of needed data usually appears to the sponsor as a barrier to the commercial introduction of a new chemical, it can only represent the tip of an iceberg of knowledge to be gained about that material. Following wide market acceptance and use, the full breadth and depth of pertinent information on a particular compound begin to be developed. Here the scientific communities have a major responsibility but one difficult to fulfill.

New drugs are typically the most thoroughly monitored chemicals because they are usually prescription drugs used under the direct supervision of a physician. The thoughtful, careful, and thorough practitioner who keeps careful records on his patients and takes time to compare notes among individuals for whom he prescribes a particular new drug may be the first to observe idiosyncrasies not seen in the clinical trials that preceded market introduction. Prompt reports of such observations (repeated if clinically and ethically feasible) to the commercial sponsor and the Food and Drug Administration may stimulate parallel observations by others and lead to a significant increase in the knowledge and understanding of the drug's properties. In addition, publication of these clinical reports in the medical literature will reach practitioners throughout the world, adding to their understanding of the usefulness and limits of the drug.

Food additives and pesticides are much more difficult to follow in public use, and most other substances even more so. Scientific and clinical serendipity often play a central role, and the most careful rechecking and detailed evaluation is then needed to affirm, modify, or negate a specific hypothesis.

Here the scientific communities play an essential role, helping to assess whether initial suspicions and technical reports were founded on good data and reasonable evaluation. Without such peer review, the likelihood is great that unconfirmed hypotheses may gain wide public credence without meeting accepted standards for scientific and/or clinical substantiation. Under these circumstances, the strength of public opinion may pressure a government decision maker and perhaps force upon him a decision in no way justified by the scientific and clinical evidence before him. To date, this has occurred infrequently, but it is a possibility to guard against.

Additional Information on New or Existing Chemicals At times, the decision maker feels compelled to obtain additional clinical or technical information, or both, before reaching his final conclusions. In these circumstances, the professional societies and the contracting laboratories provide especially important services. They direct the decision maker to individuals and laboratories with the particular expertise required and provide the facilities and scheduling flexibility essential to the prompt and timely development of new data. As soon as the new information is available, it is considered along with the results of previous studies in leading the responsible official to his final conclusions and decision.

Expanded Roles for the Scientific Communities

It is characteristic of the scientific communities that they are frequently pleased to respond to requests for assistance, but modest in making their professional expertise more widely known. In addition to responding to requests, it is important that knowledgeable professionals increasingly volunteer their detailed information and expert judgment for use in decision making on chemicals in the environment.

The staffs of professional societies need to follow current developments more closely, so that society members will be alerted to opportunities to testify in person or in writing at appropriate times during the decision-making process. If feasible, professional societies may even speak on behalf of their field of special knowledge.

Industrial research scientists, physicians in private practice and others, and contracting research and development laboratories must all be prepared to volunteer information so that decisions are founded on the best judgment and the maximum quantity of useful data. For this to happen, access to the "data-sharing station" must be easy and uncomplicated for all participants, and the use of their time must be efficiently and effectively organized.

Where data and full details are readily prepared in written form and understood, all that is needed is continuing encouragement of all qualified professionals to spend the time and effort needed to prepare their materials. But where individual participation in open meetings, public hearings, or quasi-judicial proceedings is essential, the careful planning of daily and hourly schedules becomes an absolute requisite. Otherwise, the broader inclusion of members of the scientific communities cannot and will not be feasible. The continuing responsibilities of these individuals preclude their waiting about to present their data, views, and best judgments on the issues at hand.

Concluding Comments

The scientific communities now play three major roles in the process of decision making on chemicals in the environment:

1. development of scientific, technical, and clinical data on new and existing compounds;
2. independent evaluation of these data by qualified experts; and
3. reevaluation of previous decisions in light of new laboratory and/or clinical techniques and observations.

With positive encouragement from the decision makers, members of various scientific communities can and will play a more positive role at several earlier stages of the decision-making process. To obtain the benefits of their fuller participation, the conditions under which their data and judgments are elicited must be carefully planned and effectively scheduled.

But the most important substantial new role for the scientific communities is that of educator of the public. Only with broader and deeper understanding of the basic quantitative nature of toxicity will the majority of the citizenry gain a better grasp of the nature and problems of decision making on chemicals in the environment.

PUBLIC INTEREST GROUPS

The Current Situation

In recent years, many of the efforts of EPA to regulate or restrict the use of certain toxic chemicals have been catalyzed by environmental groups. These groups, with generally small full-time professional staffs, are usually termed public interest groups. Typically, though not always, they are tax-exempt organizations. This precludes their engaging in significant legislative activity, but it does not restrict them from efforts at public education, administrative lobbying, participation in formal rule-making proceedings, litigation, and similar nonlegislative activities.

The activities of these organizations have been of two types: (1) helping to create public concern about a toxic substance through public information, and (2) directly attempting to influence the analysis and decisions of EPA on new steps to regulate or restrict a substance. It is worthwhile to consider both in depth.

First, regarding the creation of general public concern, EPA personnel have pointed out that EPA has often been forced to take regulatory action on a substance far more quickly when widespread public concern has been present. Public interest groups, using a variety of techniques, have played a major role in generating such public concern. Experience to date shows that the creation of widespread public concern which has affected decisions, while not a scientific effort per se, has in fact been based on scientific evidence; if there is no such substantive basis (as was the situation when lead compounds were first introduced into gasoline in the late 1920s), then the attempt to create effective widespread public concern is doomed to failure. As a subsidiary point, typically (but not always), the evidentiary basis is drawn from scientific evidence gathered over a long period of time, far longer than the course of the dispute itself. (The decision to ban DDT provides an illustration of this rule. The action to restrict the widespread use of NTA, on the other hand, was based on scientific evidence developed more or less simultaneously with the regulatory decision.)

From one point of view, the regulatory action brought about in part by such public concern must be seen generally as a "catch up" or reactive phenomenon to modify past decisions, and not one in which public interest groups have prevented the introduction or extensive use of new chemicals. In addition, this first function of public interest groups can be seen as creating a climate of broad public opinion such that the regulatory agency finds it must do *something* rather than *nothing;* in other words, the "cost" to the agency of doing nothing exceeds the "cost" of doing something to deal with the issue. However, this climate of opinion does not help the agency in deciding precisely what to do, and the cost to the agency may be vastly different from the cost to society.

The second role of the public interest organizations is more subtle and sophisticated. This is to suggest to the agency precisely and in detail what it can do to deal with the issue; that is, to help define for the agency what its options are in regulating or restricting the use of a substance. In recent years public interest groups have greatly increased the professional expertise of their full-time staffs; it is now quite common to find physicians, scientists, engineers, and economists as well as lawyers employed by such groups. In addition, by their active participation in the decision-making process, public interest groups can insist that the agency seek out and fully evaluate the data and suggestions from professionals outside the agency (academic scientists, for example) that the agency might otherwise tend to ignore, or who would otherwise not have access to the decision-making process. In short, the public interest groups provide both focused and timely information drawn from their own expertise and a conduit for other professionals outside government who have information pertinent to the issue at hand.

This kind of interaction has, especially in recent years, added a new dimension to the decision-making process regarding toxic chemicals. Detailed analyses of available regulatory options and their pros and cons are often undertaken collaboratively by professionals within the regulatory agency and in the public interest organizations. (This is analogous to the more partisan activity of professionals in vested or special interest groups working with professionals in the regulatory agency.) This type of participation is not well-recognized by people outside the process, since it is usually not easily communicated to noninvolved people and thus is not "newsworthy." But it is an increasingly important and effective activity of public interest groups. (In passing, perhaps we should note that litigation may be needed to establish this second function between a public interest group and the regulatory agency.)

Suggestions for the Future

The description above briefly outlines how public interest groups have participated in the decision-making process regarding toxic substances. It should be clear that this is essentially reactive behavior. It is based on the development of scientific data regarding unintended adverse effects, and equally important, it must be carried out in the fundamentally reactive mode established by the current legislation related to toxic substances. (EPA is itself in large part in a purely reactive posture because of the existing legislative framework. This is one of the main reasons why attempts have been made to develop a more comprehensive legislative framework for the evaluation and regulation of toxic substances.)

If the participation of public interest groups has made a positive contribution, it is appropriate to consider additional steps that should be taken, even in the absence of new federal legislation, to enhance and make more productive the participation of such groups in the process of regulating toxic chemicals. In the section on recommendations, we suggest several ways in which EPA can change its practices to facilitate earlier, more timely, and more effective involvement by all of those we have termed the "nonfederal publics." Public interest groups, as one of these publics, should, of course, receive equitable treatment when EPA undertakes any such initiatives.

There are, however, certain characteristics of public interest groups which, while often not unique to them, deserve to be singled out for attention in this context.

First, public interest groups are often staffed with only a small number of professionals qualified in this general area. Thus their ability to divert resources to analyze a situation quickly is limited. Many of them also often rely on a widely scattered body of scientific advisors for a major part of their analyses of, e.g., proposed regulations. Both of these characteristics suggest that public interest groups need to be informed as early as possible about proposed regulatory initiatives, provided materials for review as early as possible, and so forth.

Second, public interest groups, dependent as they are on voluntary contributions, foundation grants, and similar sources of funding, are characteristically short of funds. Thus their ability to send staff members to distant cities for public hearings or to meet with the regulatory agency staff, is seriously constrained. EPA should be sensitive to this problem and should, to the greatest extent feasible, conduct its consultations with public interest groups in such a way that financial limitations do not prohibit the full participation of the public interest community in the decision-making process. (The same sensitivity should motivate EPA's dealing with other resource-limited groups such as small business, independent unions, and academic scientists.)

Third, because of resource limitations in the public interest sector, EPA should make publications, copies of proposed rules, and other materials available at the lowest possible cost. (This is a subset of the general need to improve, continuously evaluate, and promote public information activities, which is discussed elsewhere in this panel report.)

Finally, EPA should continue to consult with representatives of the public interest sector and should consider them, as well as representatives of other "publics," for appointment to ad hoc or formal advisory and review panels.

STATE AND LOCAL GOVERNMENT

The Current Situation

The clearest involvement of state and local governmental units with the problem of toxic chemicals in the environment relates to episodic accidental spills or releases, such as transportation accidents. Dealing with such situations is often more of a public emergency than a regulatory matter. Prevention and minimization of such mishaps is the ultimate goal, but preventive activities and emergency response capabilities are beyond the scope of this study.

We note, however, the continuing need for advance contingency plans for dealing with accidental spills to be carefully and thoroughly coordinated between all relevant local, state, and federal agencies. These contingency plans should be subject to regular review so that changes in the nature of chemicals moved, the transportation routes and methods used, and new developments in emergency control methods can be incorporated into such plans.

The general trend of environmental legislation and regulatory activities in the toxic substances area is clearly away from the direct involvement of state and local units of government and predominantly toward a preeminent federal role. Given the general nature of this issue, we agree that this overall trend is desirable.

Nonetheless, certain regulatory activities are or may be conducted more effectively at the state or local level. Many states have laws and regulations governing the use of pesticides. Although the Federal Insecticide, Fungicide and Rodenticide Act provides national requirements for the regulation of such substances, several states have established more restrictive standards. In theory, under both the Clean Air Act and the Federal Water Pollution Control Act, states could regulate or otherwise restrict the use or emission of a far wider variety of toxic substances than are currently regulated by EPA. In practice, however, this is rarely done at the state level and almost never at the local level.

Suggestions for the Future

The minimal role to date of state and local governments in the regulation of toxic chemicals should not in any way be considered the optimal role. The more open decision-making process which we advocate can be effectively complemented by the capabilities of state and local government agencies in many different ways.

First, the existing governmental units which deal with public health, environmental, and industrial activities are a potentially valuable but poorly used resource for EPA. For example, state and local public health agencies can—if given timely information and, perhaps, financial aid—greatly enhance EPA's general and specific informational activities, both the activities by which EPA solicits information and those by which EPA transmits information. State and local health officers, for example, could help EPA carry out an expanded educational role, and this would be a great multiplier of EPA's own efforts.

Second, many public health professionals in state and local government are knowledgeable regarding the medical and scientific aspects of one or more toxic substances. Indirectly, they can identify others (such as academic or industrial scientists or physicians) who can also provide useful information to EPA on the selection of priorities as well as the working out of decisions on specific substances.

Third, state and local environmental agencies are often well staffed with environmental toxicologists and related professionals whose expertise can usefully be brought to bear on EPA's overall priority selection as well as on its discrete decision-making processes.

Fourth, state environmental agencies as well as state departments of commerce and industry will often know in great detail which industrial facilities in a state might be affected by the potential regulation of a specific substance. Such agencies could be particularly helpful to EPA in identifying the small- and medium-size companies that would be affected by such steps and that might otherwise be very difficult for EPA to locate. Similarly, such agencies can help EPA reach out to inform these concerns about the development of new regulations which may affect them.

These and similar steps, if properly encouraged and supported by EPA, would go far toward developing a consensus and an understanding of any decisions in this area reached by EPA. In addition, by greatly facilitating the knowledge of such decisions among affected groups, the active involvement of state and local government units at appropriate steps in the decision-making process will help speed the introduction of changes in processes or patterns of use, the installation of control technology, and the related "real world" steps that must be taken in the private sector if the goal of most regulatory decisions is to be achieved.

RECOMMENDATIONS FOR CONSIDERATION BY THE COMMITTEE

1. Consideration of Creation of an Independent Commission

In the course of its deliberations, the Panel on Nonfederal Participants in Regulatory Decision Making gave extensive consideration to the desirability of developing a federal organizational design which could most effectively respond to nonfederal participants. We recommend that thoughtful consideration be given to the creation of an independent Commission on Toxic Substances, which we feel is most likely to achieve this objective.

The commission we envision would be comprised of five commissioners, qualified by training and experience, with carefully delimited responsibilities for reaching decisions *only* on standards for such things as use patterns and emission rates, tolerances (human and environmental), labeling and other informational requirements, and prohibitions.

A particular advantage of such a commission format is that its membership could be large enough to span several of the disciplines essential to making informed judgments, while small enough to proceed expeditiously with its work. Simultaneously, its responsibilities and related authorities could be tightly circumscribed.

Clearly, the creation of such a commission would depend on congressional approval of the proposed new authorities and responsibilities, and appropriation of funds adequate to assure its effective operation. The commission would have neither research nor regulatory enforcement responsibilities, and would rely on a small staff (and on other government agencies) for data collection and organization.

2. Consideration of Creation of a Professional Environmental Education Agency: An Analogue to the Agricultural Extension Agent

Above and beyond its relations with the media, the government should take yet another important step to assure the widest possible dissemination of accurate, objective information to the public, especially to those segments of the public difficult to reach by conventional techniques. These include small businesses, consumer interest groups, people living in more remote areas, and those persons whose skepticism and suspicions about the government in general must be allayed.

A new system which combines the analagous features of a health delivery outreach program, a health education scheme, and the concept of an agricultural extension agent should be considered in formulating a professional environmental education plan for EPA. In so doing, EPA may wish to consider joint ventures with other government agencies, such as the Department of Health, Education, and Welfare, the Department of Agriculture, the Consumer Product Safety Commission, and others with similar interests and goals. Such a new program would clearly require professional training for individuals who are few in number and have gained their expertise largely as a result of experience. Thus, this program must be considered as a possible long-range objective.

Only the government is in a position to undertake the kind of responsibility suggested here. In carrying it out, however, the government must be careful to avoid any suggestion of self-aggrandizement or advocacy.

3. Positive Public Advocacy

Human rights arise out of need. The rights of individuals, peer groups, and peer group leaders to participate in a decison-making process that would control the hazards and benefits of toxic substances must be exercised if social and political decisions are to be made equitably, and if accountability is to have meaning.

Administrative procedures, implementation of "freedom of information" legislation, and the judicial process provide opportunities. They are not adequate in themselves to insure fruitful participation, and neither is merely increasing the volume of educational materials and the number of informational opportunities for the press and public. These techniques fail to identify and coordinate the shifting interests and concerns found in our heterogeneously structured culture.

Although the right of government to manipulate political pressures by manipulating the anxieties of the various publics is correctly questioned, there should be no question of the duty of government to go beyond the public notice and the press release to identify and assist social structures in coping with technological impacts.

It is necessary to specially train field workers for this task. They must not themselves "skew" the information, arbitrarily favor one group or another, or become organizers of citizen action. Rather they must find peer leaders of every kind who have an interest in specific issues, provide opportunities in training for the citizen role, and then maintain a flow of information about the facts, issues, and opportunities for participation.

The object of such programs, where they have been successful, has not been to train scientists, but rather to train knowledgeable citizens who are able to grapple with social issues and who have not previously chosen to participate in decision making in a technocracy. Without this kind of program, decision making must be relegated to the experts alone. The now extinct "community support" program of the former National Air Pollution Control Administration could serve as a model for such a program.

Labor groups, environmental organizations, social clubs, small businessmen, farmers, and ad hoc associations are examples of structures needing assistance in the exercise of their rights.

4. Routine Review of Research and Development Activities

As a corollary to the formation of an independent Commission on Toxic Substances, we propose that the research and development programs undertaken by EPA be routinely reviewed by a panel of qualified experts. The scientific, technical, and clinical relevance, and the regulatory pertinence of each research project proposed for funding—whether to be performed in-house or by a contract research and development laboratory—would be reviewed, in advance, to assure that it is appropriate to achievement of the agency's mandates and goals, and again, at its conclusion, to evaluate how effective and timely the results have proven to be in light of the initial purposes and schedule. The findings of this review group should be published at least annually, so that all interested persons may be informed.

APPENDIX E

Working Paper on Equity

SUMMARY

Problems of equity arise when the risks associated with introducing chemicals into the environment are not restricted to the persons who benefit from them. Inequitable distribution of costs and benefits can be seen among generations, industries, geographic areas, occupations, social and economic classes, and other components of society. These equitable considerations cannot be aggregated and incorporated simply into economic benefit–cost analyses.

Distributions of benefits and costs are altered by regulation through political and social mechanisms. To promote more equitable results of decision making, careful attention should be given to fairness in the decision-making process itself. Differences in the capabilities of interested parties to influence decisions should be lessened, and the decision maker should actively solicit the views of as many interests as possible. The parties who will be affected by the decision should be represented in proportion to the effects they will bear. The federal government should sponsor coordinated research to provide the necessary information for decision makers as early in the regulatory process as possible. This information, coupled with participation by affected parties, could so strengthen the decision scientifically and economically that reliance upon the appeals process to obtain a satisfactory and final decision would be significantly lessened.

THE CONCEPT OF EQUITY

Equity means the fair distribution of benefits and costs, but a fair distribution does not necessarily mean an equal distribution. For example, it can be argued that a project which increases inequality is justifiable on the grounds of fairness to the degree that the inequality contributes to the well-being of disadvantaged groups (Rawls 1971). On the other hand, gross inequalities often lead to the presumption of inequity.

It seems clear that if the costs and hazards of a new toxic chemical are limited to those who benefit from the chemical, the problems of equity are greatly diminished or even eliminated. This situation, however, is unlikely to arise in practice. When the distribution of potential costs is different from the distribution of benefits, the equity problem increases in importance as the difference increases.[1] Judgment of whether the redistributional effects of a toxic chemical are inequitable is fundamentally a problem of political and social choice. The proper rationale for government intervention is that some markets are inefficient and do not sufficiently internalize externalities, or that even efficient outcomes are regarded as unfair.

The government acts as an overseer of the various conflicting interests, taking into account the redistribution of costs and benefits of a specific regulatory strategy. But it is often extremely difficult to resolve the issue of fairness in distribution, or even to establish the grounds upon which such judgments could be made, by considering the possible outcomes. Therefore, attention is paid to fairness in the decision process itself, because one way to promote fairness in the outcome is to establish fairness in that process.

We offer this as an operational principle of equity (or fairness) in the decision process: the decision process is equitable when the potentially affected individuals are represented in proportion to the effects of the decision. We are concerned in particular that the persons who will bear the hidden costs of toxic substances and the members of future generations be represented. In practice, we will never fully know the potential distribution of costs and benefits, so we will not be able to precisely match representation with distribution of effect. But the process of trying to establish the distribution of effects will generate

[1]If potential redistributions from use of a chemical are small, they often are neglected in practice, and the overall benefits and costs are simply treated as in the analysis of economic efficiency.

valuable information and promote informed debate; indeed, this is a key feature of the decision process.

If the nation is to be served best, environmental decisions should be balanced among the large number of interests to be affected. Consideration of all interests can increase the likelihood of an economically efficient outcome (i.e., maximization of net benefits). In recent years a trend has developed to narrow the scope of the decision-making process as a way of protecting innocent third parties (e.g., the Delaney Amendment).[2] Narrow rules might be the most practical way to protect people from diffuse environmental risk, but if these people are fairly represented in the decision process—including representation by proxy for the unborn—the need for narrow rules is reduced. Moreover, there should be concerted effort to determine the effects and interrelationships of anticipated environmental regulatory actions with other governmental programs and activities.

SOME EXAMPLES

A new pesticide is produced that can increase crop yields by one billion dollars annually for five years until insect tolerance makes it ineffective. The pesticide also has a 10 percent chance of increasing intestinal cancer by 40 percent 20 years after application. Considered strictly as an efficiency question, one would discount the expected costs of mortality and compare them with the present value of the benefits. For most efficiency (benefit–cost) analyses, the distribution of benefits and costs is not of concern. Attention is given only to the net benefits after discounting future benefits and costs to present values. Considered as an equity question, however, one must ask further whether it is fair that "we" get the benefits and "they" get the costs.

People who live near industrial areas comprise a lower economic group than those who purchase most of the industrial goods. By virtue of their proximity to the industrial operation, those in this lower economic group are exposed to a higher level of pollution. A relevant question is whether it is fair that an initial distribution of wealth should determine the extent to which one is exposed to pollution.

Workers bear a disproportionately high risk of exposure to toxic materials, since the source of pollution is the manufacturing sector, and

[2]This clause is in the Food Additives Amendments of 1958; it provides in part that "no additive shall be deemed to be safe if it is found to induce cancer when ingested by man or animal, or if it is found, after tests which are appropriate for the evaluation of the safety of food additives, to induce cancer in man or animal. . . . 21 U.S.C. sec. 348(c)(3)(A) (1970).

pollution within manufacturing facilities frequently is more severe than that outside. Those workers who live near the factory suffer "double jeopardy," since both their work and nonwork environments are worse than average. A relevant question is whether workers should bear the bulk of the risk of harm for the society that benefits from the product of their work. It is often argued that equity is not an issue, because workers have the choice of accepting higher pay for more hazardous jobs. However, because the market fails to function perfectly, perhaps because the workers do not know the nature of the risks or because they lack mobility and alternative jobs, inefficiencies in the market can lead to important inequities.

Stringent legislation to control toxic substances may affect smaller firms more than larger ones, since the smaller firms often are economically marginal and are not able to take advantage of the economies of scale or of resident expertise in testing materials. They have inadequate reserves to cover both the new and continuing costs of compliance and generally operate with shorter time and money budgets. Fairness among firms and among their owners and employees is of appropriate concern.

Some industries or regions of the country might be affected more by regulation than are others, and the regulation might on occasion increase unemployment in some areas. Also, the increased cost of domestic production caused by regulation might shift production abroad. Fairness in these transfers of labor and capital should be considered.

While primarily intended to preserve or improve environmental quality, the regulatory process may provide or be accompanied by changes in distribution of wealth or other resources. Further, the pattern of internalizing those environmental costs that previously were not included may lead to the accrual of costs to parties other than those who enjoy the benefits. It is not clear to what extent society may judge this to be equitable, but we recognize that questions of equity usually are resolved through the political process.

The most that economists and scientists can hope to do is to reduce uncertainty and elucidate the nature of the underlying trade-offs which must be made. If the decision-making process is equally accessible to all affected parties, if minority interests are not "weighted away," and if the regulator is held accountable for stating explicitly the trade-offs to be made, more responsive and responsible environmental decisions may be made.

The distributional aspects of the regulatory decision are of prime concern. The central question is who bears the risk of harm and who is likely to reap the benefits of a particular regulatory strategy. There must be concern for distributional implications, not only between manufacturers and users, but also among (1) individuals (e.g., rich vs. poor), (2) groups of citizens (e.g., workers and consumers or consumers and taxpayers), (3) firms (e.g., large vs. small), (4) industries (e.g., chemical vs. textile), (5) geographic regions of the country, and (6) international interests.

It is important to realize that the equity issues ultimately concern individuals. Firms bear the organizational costs and act in the interest of the individuals and, therefore, have a place in the decision process.

It is not easy to balance the welfare of workers and consumers or of large firms and small firms. However, the decisions ought to be made explicitly so that the political process can operate in the open and as fairly as possible. Furthermore, once the decision is made, the regulatory strategy must be fairly implemented. The transient effects can be reduced by proper timing of compliance and by multistage standard setting. This is important because the abruptness of change caused by some decisions is as important as the intended change in conditions.

ACHIEVING EQUITY DURING AND AFTER THE DECISION

Those concerned with equity in regulatory decisions will want to ensure that both the process of making a decision and the results of the decision are fair. The regulatory process consists of gathering and processing information, establishing direct control mechanisms (e.g., standards and permits), enforcing the mechanisms (e.g., inspection and monitoring), using indirect control mechanisms (e.g., tax incentives and use charges), and resolving judicial and quasi-judicial procedural issues (e.g., allocation of the burden of proof).

During the various stages of the decision-making process, equity requires including all of the relevant factors in that process and representing the affected parties so that the distributional implications may be considered properly. Deciding which factors are relevant is a key issue and can reflect in part the value judgment of the decision maker. At a minimum, however, the following activities must be undertaken: (1) internalization of the monetarily quantifiable externalities in the benefit–cost statement (e.g., lost wages), (2) delineation separately of the difficult-to-quantify health and environmental effects, not necessarily in monetary terms (e.g., pain and suffering), and (3) understanding of differences between benefits and costs at different times and among different segments of the population. Quantification, monetization, and aggregation are simplifications that often obscure the important distributional considerations.

DISTRIBUTION OF RESOURCES

Differences in the ability of the various parties interested in a pending regulation to gather, use, and inject information into the decision-making process often prevent fair representation of all sides of the issue. The producer sector of the economy generally is better equipped to protect its interests than is the consumer sector. The organizational costs here are not assumed by firms to the same extent they assume the benefits (Olson 1971). Because in this process information is power, the present imbalance could lead to both inequitable and inefficient regulatory decisions.

Information must be judged on its quality, relevance, and reliability, taking into account the possible biases from its source. Technical expertise usually has been available to the chemical producers, but other parties frequently lack this resource. Also, the parties are not fairly represented on advisory committees and other organizations relied upon or consulted by government. This reality is likely to produce an unbalanced and unobjective assessment of the benefits and costs. Improving the balance of resources among interested parties often would improve the usefulness of information available to the decision maker. At present, the parties might tend to exploit competitive advantages in resources—scare tactics of some environmentalists versus self-serving research studies and advertising of industry. With properly balanced resources, the issues could be addressed more directly and with greater clarity. In brief, sound information is a scarce resource.

BURDEN OF PROOF

In making decisions with incomplete knowledge of the consequences, we must proceed with great caution to protect the interests of future generations. Until relatively recently in our industrial development, the effects of any one decision were of rather short-range consequence. Decisions today have much greater potential consequences, and many of them might preclude options for the future. Should it later be determined that we were more conservative than necessary, we would retain the ability to modify the decision at relatively small cost. If, on the other hand, we err in the opposite direction, it may be extraordinarily difficult or even impossible to reverse the effects of the earlier decision, thus passing on great costs. It is because the decisions today are so momentous that the equitable considerations are so important.

Equity, unlike efficiency (which tends to dominate equity in quantitative analyses), cannot be reduced to a single aggregate value. Subjective considerations involving equity must be given additional attention in establishing policy for distribution of chemicals in the environment if we are to strike the proper balance of distributing both benefits and costs among generations.

A central equitable issue is how the burden of proof should be allocated among parties to the decision. In the early days of our industrial revolution, the growth of industry was regarded as a hallmark of progress, and its desirability usually went unquestioned. As our industrial capabilities and our reliance upon technology increased, the long-range consequences of technological actions became such that any one action could have disproportionately large effects upon other components of society. On whom should be placed the burden of convincing the decision maker that the effects would be either reasonable or unreasonable?

The trend now has shifted from acquiescence by society in the acts of individuals toward placement of the burden of proof on the individuals who propose to carry out projects with probable environmental consequences. This is an appropriate and healthy trend, because the entity that proposes the project usually is the one that plans to benefit from it and usually is the one with the resources to analyze its unfavorable as well as its favorable consequences. The government should not be obligated in all instances to ascertain if existing products are harmful when the producers could have been required before distribution to provide assurances of safety and freedom from environmental hazard.

RECOMMENDATIONS FOR CONSIDERATION BY THE COMMITTEE

1. The federal government should more actively encourage coordinated research to gain information needed for regulatory decisions. A particularly challenging problem concerning the burden of proof relates to the review of chemicals already on the market for which decisions were made at an earlier time. The challenge is to incorporate new scientific evidence into past decisions. In many cases, this implies some additional research, or at least new interpretation of knowledge. In the specific case of pesticides, the law provides that pesticide registrations are automatically canceled every five years and are renewed only after the manufacturer has satisfied the government that its product continues to meet certain criteria (Federal Insecticide, Fungicide and Rodenticide

Act, 7 U.S.C. sec. 135 et seq. [1975 Supp]). In practice, the process of periodic review is not performed well, and scientific review is often perfunctory. Accordingly, the regulatory agency should assume a firm responsibility to support research where appropriate to bring new science to bear on old decisions.

2. The details of decision making should be made public. This can be done through the publication by the Administrator of a white paper for each important regulatory action undertaken. It should include the details of all considerations—economic, legal, scientific, and others—taken into account in reaching the decision and should demonstrate an acknowledgment of the diverse points of view offered by the public, by the public's representatives, and by other government agencies. This paper should outline the available avenues for the decision and discuss the reasoning followed by the agency in selecting the one it did. It should be issued when the Administrator announces a decision to act concerning releases of chemicals into the environment. We foresee this as being similar in purpose to the environmental impact statements under the National Environmental Policy Act of 1969 (42 U.S.C. sec. 4321 et seq., 1970). The Administrator should explain in this paper how he decided whether NEPA required an environmental impact statement for the decision.

3. The initial decision process for which the administrative agency is responsible should be so strengthened scientifically and economically that it will not longer be assumed that the ultimate issue will be decided only when it reaches the appeal process. Appeal and adjudication of regulatory actions have become increasingly important in the overall decision-making process in recent years, and a tendency has developed to consider these appeal mechanisms as the principal opportunities for bringing forth opposing points of view and for arriving at decisions through adversary proceedings. The trend toward reliance on the appeals process is unfortunate. As a result of this trend, the Administrator's early decisions in some cases become almost academic. Opportunities for public input—even through active solicitation—from all interested parties should be increased at an early stage in the regulatory process. To this end, the provision for a predecision public hearing allowed by the pesticides legislation should be extended to other regulatory areas. The Administrator should consider how prehearing exchange of information among parties could be facilitated through depositions, interrogatories, discovery, and other procedures.

4. The present imbalance in the information and resources brought to bear on the benefits and costs should be redressed in one or more of the following ways:

a. All interested parties should be fairly represented on advisory committees and other bodies that make substantive or procedural recommendations to the agency.

b. The regulatory agency could itself become an advocate of the environment. This is not too far removed from the current situation, but the information-gathering function should be separate from the standard-setting function, somewhat similar to the relationship between the National Institute for Occupational Safety and Health and the Occupational Safety and Health Administration.

c. The agency could create a more frankly adversary process and a more evenly matched one. It could contract with independent laboratories to develop only the cost estimates, continuing to rely upon industry to develop the benefits. The hearing process could be structured as a well- defined adversary process. Over time, the regulatory agency would create "environmental prosecutors" to match the currently available "chemical defenders."

d. Some other agency, for example, a Consumer Protection Agency, could assume the responsibility of arguing the overall environmental interests and thereby balance the adversary process.

The purposes in creating environmental advocates are to develop higher quality information in the advisory process, to lessen the current burden upon the agency of representing the otherwise unrepresented interests (including future generations), and to prevent "industry capture" of the agency over the years.

5. Any information available to the Administrator on the hazards of a chemical that is regulated by the agency should be available for public inspection in timely fashion during and after the regulatory process. The Administrator should make public the agency's procedures for implementing the Freedom of Information Act (5 U.S.C. sec. 552, 1970), with emphasis given to any distinction between how publicly and privately acquired information is treated.

REFERENCES

Olson, M. (1971) Logic of Collective Action: Public Goods and the Theory of Groups. Cambridge: Harvard University Press.

Rawls, J. (1971) A Theory of Justice. Cambridge: Harvard University Press.

APPENDIX F

Working Paper on Governmental Information Needs: Benefits

WHO BENEFITS FROM CHEMICALS?

One of the clear mandates of the U.S. Environmental Protection Agency (EPA) is to expand the net benefits of chemicals with respect to man and the environment. Yet of the various benefits discussed in this paper all are to man and none to the environment. This is so because only human beings testify as to benefits. Some humans may testify on behalf of the wolf species, while others will testify on behalf of the sheep. Some humans will testify on behalf of trees and natural wonders, while others testify on behalf of cheaper lumber and better housing for humans. In all cases they can speak only to human values with respect to nonhumans or inanimate objects. The latter cannot tell us what they consider benefits to themselves.

We note this obvious fact to emphasize that EPA, like other organizations, can never act on behalf of the environment, but only on what humans urge be done for presumed benefit to the environment. EPA, again like other organizations, must trade off different and often conflicting sets of human values in making its decisions.

WHAT ARE THE BENEFITS FROM CHEMICALS?

The known benefits from chemicals are those perceived by human beings, who may concern themselves with benefits to individuals, to mankind as a whole, to other species, or to the inanimate physical world.

Benefits that have been so perceived cover an enormous range, for human goals and satisfactions vary greatly. Many of these are summarized below.

ADVANCING HUMAN HEALTH

By Extending Life

Soap and chlorine provide cleaner skins, garments, and water (for drinking and bathing). Chemicals that render garments flame retardant extend many lives. Since no life can be continued forever, it is useful to speak in terms of "extended" life. The term also emphasizes the fact that "saving" the life of a 10- year-old child could add more years of life than "saving" the life of a 70-year-old man (which tells us nothing, of course, about the desirability of saving different types of lives).

By Reducing Suffering and Pain

Vitamins can remove the scourges of scurvy and pellagra; DDT ended malaria. They have significantly reduced the years of pain and low vitality that had previously afflicted persons with such diseases. Chemicals have reduced coccidiosis among chickens, have increased beef yield, and have otherwise made meat cheaper and more available. This in turn has reduced the incidence of some illnesses caused by protein deficiencies. Chemicals for cleaning dairy barns similarly help keep down the price of milk, foster its use, and thus reduce nutritional diseases.

REDUCING HUMAN WORK EFFORT

Gasoline is an example of a chemical that reduces human effort: an alternative open to society in place of gasoline-powered engines would be to use railroads more extensively, but this would require additional work in coal mines to provide equal transportation service on those lines where diesel fuel was not used. For farmers the alternative to the tractor is the horse, requiring considerably more care and more work to grow, cut, store, and feed hay.

Butyl rubber and asbestos permit better insulation against heat loss and thereby reduce the work required to mine fossil fuels or to grow timber to compensate for heat lost without insulators. These examples, chosen intentionally, deal with chemicals that involve important questions of health.

Gasoline, polyvinyl chloride, aluminum, and a host of other chemicals are used that make it easier to provide equipment for hikers and hunters to leave crowded cities and enjoy recreation under natural conditions.

NATIONAL SECURITY

Uranium hexafluoride is manufactured for weapons. Airplane fuel is used for military aircraft. Still other chemicals are used indirectly in the manufacture of uniforms, military installations, and so on.

CONSERVING RESOURCES

By consuming less lumber and iron today, we could conserve these resources for future generations. But we do not wish to deprive ourselves today on behalf of future generations, and we therefore use chemicals so that we can consume resources more efficiently. Thus, rust preventives enable us to consume less iron, and DDT spray allows us to use forests more efficiently (by killing moths that attack trees).

CONSERVING IRREPLACEABLE OBJECTS FOR FUTURE GENERATIONS

To conserve trees all we need do is consume fewer of them. But conserving one-of-a-kind objects like the Mona Lisa or the Grand Canyon is quite a different matter: it is, for practical purposes, impossible for a given generation to replace these. The chemicals developed to preserve priceless paintings or manuscripts provide uniquely valuable benefits. So do those that slow natural erosion in the streams flowing into the Colorado River, thereby slowing the eventual destruction of the Grand Canyon.

HUMAN CONVENIENCE

The propellant material used in spray cans provides convenience to the user. But aerosol cans of paint, for example, cost considerably more than older methods for applying color. Hence, consumers have demonstrated—by the margin of that increased payment—that they value the increased convenience. Convenience also appears in the use of motorcycles or bicycles—whether to reach local destinations or more remote areas like national parks which could also be reached less conveniently on foot. Hikers' boots, too, now have chemical composition soles that grip the ground more effectively and conveniently than do leather boots—at least they do so better than leather at any equivalent cost to the hiker.

Many human recreational activities involve convenience—nylon tents replacing canvas ones; plastic marine hardware and fiber-glass hulls replacing older materials that demanded endless maintenance, and so on. A vast variety of consumer products involve convenience, for example, for the housewife (who can clean ovens in a tenth the time with new products), and for the parent (who can buy a plastic doll house instead of building one with 10 times greater effort).

AESTHETICS

Paints, deodorants, synthetic food flavors, lipstick colors, textile dyes—all create aesthetically pleasing results, as judged by those who buy the products, and, at times, by others as well.

IS THERE AN OBJECTIVE, SCIENTIFIC WAY OF MEASURING TOTAL BENEFITS?

No. Nor is it clear whether there will ever be such a method. For until all human beings in a society are so homogenized that their purposes are identical, their goals and attitudes and tastes the same, they will not give the same priority to their values. Nor will they adhere to each value with equal intensity in practice. And thus there is exceedingly little likelihood that there will be an objective way of summarizing all those values for a benefit–cost calculation.[1]

Yet EPA, by the Acts of Congress under which it is required to operate, is committed to precisely this summarization of all benefits as judged by human values, and then to a similar summarization with respect to all

[1]It has been suggested that asking questions of consumers in surveys would establish their "willingness to pay," and from those answers one could infer the "consumer surplus" that consumers are enjoying. We know, for example, that if the prices of lifesaving medicines were higher, some of the buyers of such medicines would surely pay more than they now have to. However, a vast body of survey experience indicates that questioning people does not produce infallibly reliable answers. And a large body of psychological analysis indicates that the more respondents are aware of the importance to them of varying answers, the less certainly an outsider can rely on the truth of their verbal responses. In any event, a measure of consumer surplus would be of no use to an administrator unless he had such a measure both for the chemical in use and, necessarily, for every chemical that might be substituted for it if it were either ruled out altogether or limited in different use patterns. The resultant sequence of hypothetical figures would lack the objectivity that an administrator requires even if there were no conceptual problems involved.

costs—all this as mere preliminaries to judging the net advantage of using a given chemical.

Lacking a more objective method, then, EPA must inevitably give substantial weight to the summarizing of net benefits reported in the dollar sales total for those who buy particular chemicals.[2] It would be well if Congress were to establish guidelines for valuing benefits. EPA could then work at the less ambitious but still herculean task of establishing the facts in particular cases, and then reaching decisions that accord with the value priorities that Congress had set forth. The result would more appropriately implement the purposes that EPA exists to carry out.

WHAT BENEFITS MUST EPA CONSIDER?

IN PRINCIPLE

We are agreed that EPA will usefully consider every class of benefits that is treated as significant by buyers of chemicals. In some instances, a particular class of benefits will obviously be trivial and will receive little attention. The same conclusion will follow for some specific uses.

But EPA must consider benefits in addition to those that private consumers recognize. As with costs, some may have been noted by consumers. They may not have been recognized or reasonably weighed by them. Still others may not interest existing buyers. EPA's mandate, however, does not permit it to ignore any class of benefits to which any group of serious spokesmen wish to call attention. How extensively it will wish to consider them, or what values it will in the end decide to attach, are matters within its province, for it carries the ultimate responsibility for weighing and then deciding.

IN PRACTICE

In general it may be presumed that a private firm seeking permission to market a new product, or desiring to continue marketing an existing one, is ready, willing, and able to provide the most comprehensive list of the

[2]"Pareto optimality" has often been mentioned in connection with achieving objective judgments. It is not likely to be very helpful in the administrative process. "Pareto optimality" refers to a situation that may never face an administrator during a long and stormy career. It is a situation in which one can anticipate that action will make at least one person better off without at the same time making anyone else worse off in any respect. Given such happy premises, the action should be taken.

classes of benefits that chemical will provide. By also providing sales figures, by type of user or application, the firm will give a useful indication of the relative importance of the chemical. (For example, certain classes of users may be presumed to use the product only to expand production of a food; others, to change its market acceptability by changing aesthetic aspects; and so forth.) EPA will review that list, and those estimates, to test for reasonableness and reliability. It may do so, in some instances, by varieties of adversary procedure. Since EPA seeks a comprehensive view of benefits, it must consider whether there are other benefits not noted by the applicant. While firms generally make the best possible case for their products, a small firm, for example, may not have the time or capacity to do so. In all cases, however, EPA will have to add to that list benefits to aesthetics, national defense, and so on, though these may not be of interest to the actual purchasers of the product and hence may not be noted by the company.

HOW IS EPA TO COMPUTE BENEFITS?

The panel notes below three considerations relevant to EPA's assessment. Each of the three is suggestive at best; the three together are useful, albeit not controlling.

The first consideration is that most chemicals will continue to be bought and sold in the private market. EPA must therefore give serious attention to the prices and sales totals that appear in that market. These summarize better than any other single method the sum of expected net benefits as judged by purchasers. Since most purchases are part of an ongoing lifetime purchase pattern, rather than once-in-a-lifetime purchases, there is even a presumption that most purchasers received the benefits they expected. Those who were disappointed ceased to buy, even as new purchasers were first trying the product. On the average these expenditures measure the sum of benefits buyers expected from those purchases.

These judgments are peculiar to time and place. If consumer incomes were different (or prices, or knowledge, or government regulations, and so forth), there would undoubtedly have been different expenditures. But that succession of "if's" does not create that alternative world. Given the prevailing conditions in American society, price generally measures benefits expected by those who buy.

Second, the very existence of EPA indicates that Congress and the Executive are no longer prepared to take private market judgment of chemical purchasers as a sufficient indication of the benefits and costs of chemicals. EPA has, therefore, the responsibility of adding into the benefit listing and judgment an allowance for benefits that private purchasers may ignore. For example, by purchasing one chemical rather than another because it is more convenient to them, buyers' prices will not reflect the fact that it may be better for some social purpose that does not interest them. Thus, an agent that cleans ovens more quickly may put far less effluent into public sewage systems than one that involves repeated washing with abrasive agents. The benefits to the municipality, and eventually the improvement in water supply, are benefits that must be considered together with those that the purchaser noted. (Equivalent adjustments have to be made on the cost/hazard side as well.)

A third consideration is that EPA is actually not interested in assessing benefits versus costs (including hazards) for any chemical. Rather it is concerned with doing so for a given chemical and for the pattern of chemical use that would replace it if EPA banned the particular chemical or restricted its use. EPA must, therefore, array its best information on the benefits and costs for that chemical and for as many alternatives as it considers to be reasonable regulatory alternatives.

BENEFITS OF NEW CHEMICALS

New chemicals should be evaluated before permission to market them is given. Should such permission be given at all? Should it be restricted to certain applications or certain locations? For selecting among such alternatives EPA will, of necessity, have to use speculative projections of sales as the only ones that can be made. However, these speculations can be constrained by two elements in the real world. First, the new product is likely to replace another chemical already in use (for example, one yellow dye replaces another). Or it may replace a procedure already in use (adding chlorine to the water supply provides cleaner water, for example, replacing the older system of water carriers who bring water in bottles or casks from natural wells with good water).

As a first approximation, the chemical producer can reasonably be asked to supply what most substantial companies today supply before marketing a new product, i.e., a market projection (with a few basic figures on the price at which they expect to market the product and the quantity of sales expected over the first few years). As with toxicological data supplied by these same companies, it is hardly to be expected that the data will be precise or wholly objective. But they do constitute appropriate raw material for EPA, which can then subject the data to tests of consistency and reasonableness.

As with the toxicological data, we presume that the major constraint on the care with which such data are prepared for EPA's use will be the continuing relationship between the manufacturer and EPA. There will be other applications to be made by the manufacturer, at a later date, for another new product or use. The manufacturer may yet have to make a showing merely to keep one of its products on the market. If it engages in misstatements, it must expect that to affect its subsequent treatment. We assume, moreover, that sales projections will be assessed by EPA as carefully as actual statements of sales.

BENEFITS THAT EPA CAN IGNORE

Among the benefits of a given chemical many persons would include the jobs it provides in its production or sale. But this should not be so for EPA's purposes. Employment in producing any given chemical is at the expense of employment in producing other things that are useful and desired. So is the investment in capital that could be used to make other lines of chemicals or other products. If human beings wanted no services or goods other than a particular chemical, that conclusion would change. But since people want many goods, the human labor and investment devoted to producing any given one is necessarily made by producing less of some others. Even if we were in the happy position of producing all the goods society desired, we would still have to consider whether reducing the work day or week for some workers would not be preferable to continuing to produce a given product total.

The issue behind this proposition, however, is that closing down a plant will impose real and significant costs on particular workers, particular investors, and particular municipalities. Such are the inevitable consequences created by the desire to regulate chemical use; if there were no regulation to achieve the goals specified in the acts under which EPA operates, no such disemployment of men and resources would be created. It follows that costs of adjustment necessarily created by regulation must be considered as costs. They belong in the array of costs used when EPA compares costs and benefits to reach a decision. (Varying dimensions of regulation, from total banning to partial limitation, may reduce such costs.)

It is, of course, possible to go still further. Congress might take action to compensate workers, investors, municipalities, countries, and states for losses as a result of EPA regulation. Similar issues have been addressed in various ways by the Congress and the courts. But they cannot reasonably be addressed merely by EPA, which is only one of many federal agencies taking actions that inevitably and necessarily injure some citizens and provide bonuses to others.

Similarly, EPA actions may change the U.S. balance of payments position, the level of competition as compared to that of monopoly, and so forth. Neither those consequences considered admirable (benefits) nor miserable (costs) should be part of EPA's decision calculus. It is up to other agencies of government, and the courts, to deal with such generally marginal effects.

APPENDIX G

Working Paper on Regulatory Information Needs: Hazards and Costs

SUMMARY AND INTRODUCTION

Proper regulation of hazardous environmental chemicals is conceptually simple. Three questions must be answered of which the third poses the severest problem.

1. Is the chemical hazardous?
2. Is the chemical beneficial?

If the answers to one or both are no, the decision is straightforward. If the answer to both is yes, the third question must be asked and answered.

3. Do the benefits outweigh the hazards?

This panel considered the elements of hazards to human health and the environment and of cost related to the regulation or nonregulation of a hazardous chemical. The key initial element is the hazard evaluation, for if no hazard is identified or perceived, there is no need to go farther in the analysis. Hazard determination is a function of both the intrinsic toxicity of the chemical and the pattern of use projected for it. Many toxic effects appear to have a threshold, are reversible, and are characterized by easily identifiable dose–response curves. These we call high likelihood hazards. If toxicity is high, hazard can be controlled by relatively simple use limitation and control. Chemicals which cause irreversible effects, such as cancer or mutation, generally do not have an effective threshold or no-effect level. Typically they have only a low likelihood of causing such an effect. Yet if they do, the personal cost is high indeed. Control measures generally need to be stringent and therefore are costly per unit protected and are vigorously contested. Regulatory decision makers face the severest test in this area.

Uncertain hazards represent the black box of regulatory decision making. They are typified by second- and third-order effects and interactions with other chemicals. Action initially is rarely required or permitted on the basis of the information available, but constant surveillance is needed for clues that potentially serious effects are occurring.

The data base needed to define hazard consists of both human and environmental toxicity data and estimated or extant use patterns (NAS 1975). A sequential series of steps may be envisaged for the toxicity analysis. Class of chemical compound and results of simple screening tests may either be adequate or suggest the need for further and more extensive studies. Estimation of projected use may be difficult for new chemicals. Environmental persistence is a key factor and can make the difference between a fairly hazard-free chemical and a seriously hazardous one.

The reliability of the data base must be constantly reassessed for each individual compound and for the predictive toxicological systems themselves. The government, in cooperation with industry and academia, must play a strong role here. New efforts should be made to assess the reliability of use estimates. The total toxicological data base could be enhanced if appropriate and equitable processes were established to permit greater availability of data possessed by manufacturing companies.

A compound nonhazardous at initial introduction may become a significant hazard as new, different, and more extensive uses are devised. Methods must be developed to signal and assess such new uses.

Hazard evaluation is the critical first step in ensuring that a chemical is environmentally safe. Attention needs to be given to the share of resources allocated to the development of the science of hazard evaluation. Many persons are convinced that overall support for such efforts is seriously inadequate.

Costs are often stated in terms of dollars, human death and suffering, or environmental degradation, but these may all be considered as forgone human or environmental opportunities. We feel it is not possible to aggregate all costs into a single unit of expression. Thus the decision maker will be faced with equations containing incommensurables, but he can be aided by data value judgments, organization, and presentation.

Measures of direct nonpecuniary cost to human health and the environment flow directly out of the hazard analysis. These must be given quantification in a descriptive probabilistic manner.

Certain direct economic costs, such as transactional and control costs, can be given reasonably precise pecuniary estimates. Indirect economic costs are more difficult to estimate and require complex estimates of indirect effects. For example, the ultimate cost consequences of the need to find substitutes may have complex and far-reaching implications. Distributional costs between areas and individuals can be substantial, as, for example, when certain factories are closed and others are built or expanded. The positive or negative legacy we leave future generations must be considered.

Of critical importance but particularly difficult to define and quantify are the cumulative costs of regulatory decisions on the industrial structure and on future innovation and entrepreneurship. Data needed to assess all these costs are available from a large, often bewildering, variety of sources and present troublesome considerations related to timeliness, relevance, and precision. Special efforts are needed to assure proper estimates of probable error in cost estimates since it is particularly difficult to look back to determine the precision and accuracy of many such estimates, particularly those of indirect costs used earlier in the decision-making process. Table G-1 suggests a way in which cost data can be organized for regulatory decision making.

DEFINITIONS

Hazard: A function of the *toxicity* of a chemical and its *use.*

Toxicity: Intrinsic characteristics of any chemical. *Qualitative*—the recognizable manifestations of the interaction between the chemical

TABLE G-1 Organization of Cost Data for Decision Making

Type and Example of Costs	Pecuniary	Nonpecuniary	Units
Indirect, noneconomic–human mortality	No	Yes	Decrease in life expectancy
Direct, economic–control of toxic impurity	Yes	No	Money
Indirect, economic–change in income distribution	Yes	Yes	Money and judgment of equity
Structural–change in output concentration ratios	No	Yes	Percent of total output in 5 largest producers

and a living system. *Quantitative*—that concentration or amount of the chemical that results in a recognizable manifestation.

Use: Any opportunity for exposure, including concepts of quantity produced, pattern of distribution, and extent and duration of contact with man or the environment. This includes explicit factors of persistence and bioaccumulation. The entire breadth of the material balance display must include natural background levels for naturally occurring agents.

Cost: Forgone human and ecological opportunities.

Safety factor: Estimate of the difference between permissible population exposure levels and the highest level that would not be expected to cause hazard.

Validation: Process of comparing predicted costs with actual costs and predicted values or effects with those later observed in man or in other animals.

Reliability: An estimate, often judgmental, of the firmness or "real life" quality of biological and economic data used to estimate hazards and costs.

HAZARD ANALYSIS

CONSTRUCTION OF HAZARD RATINGS

A desirable adjunct to the general classification scheme for the categories of chemical hazards noted above would be a system for rating or ranking such hazards, especially hazards to humans as the subject at risk. Such a rating system must take into account, as a minimum set of important contributing factors, the generalized hazard H; the intrinsic human toxicity factor T peculiar to a given chemical or product; and a factor U stemming from the conditions of its usage and consequent modalities/levels of exposure of man. As a first approximation, the relationship between the hazard function H and scale factors T and U derived from intrinsic toxicity and usage may be expressed as H being a function of T and U, or $H = f(T, U)$. Implicit in this notation is the necessary condition that both T and U scale factors have a time variable component, so that single acute exposures, chronic low-level exposures, and intermittent interaction between tissues and chemicals can be accommodated by the appropriate usage functions contributing to the hazard estimations.

The toxicity function T for a given chemical would ideally be generated by specific information bearing on morbidity or mortality in humans under well-defined exposure conditions. Since such controlled clinical, epidemiological, and individual exposure information is rarely available, it will usually be necessary to supply its near equivalent in the form of a toxicity index derived from carefully designed experiments with the most suitable animal model exposed by the route of interest, such as skin contact, inhalation, or oral ingestion. Toxicological research with one or more animal models will in each instance generate precise dosage versus response relationships, on an acute or chronic basis, in close simulation of the human modalities encountered.

The greatest difficulty may reside in generation of the usage component U of the hazard equation. The factor U must accommodate temporal variation in exposure patterns; it has to represent both the probability and intensity of exposures as a distribution function covering a population of individuals at risk, and yet it must be reducible to the format of an integer scale factor in order to be readily usable in assessing magnitudes of hazard. This formidable combination of conditions imposed on formulation of a scale for U factors presents a challenging conceptual problem, but one which should be amenable to systematic solution by suitable combinations of distribution functions and temporal decay curves.

The categorization of hazard to man in the form of index values of H for a given chemical and exposure modality as represented by $H=f(T,U)$ is subject to several sets of options. In this regard it may be sufficient to note that the realities governing human exposure to a given toxic chemical may find *either* T or U as a major contributing factor to H, depending on the nature of the chemical. Therefore, in generating the hazard function H for specific classes of chemicals and exposure modalities, modeling the relationship between T and U may require the use of simple first-order functions of each variable (e.g., $H - TU$), of exponential functions (e.g., $H = TU$), or of linear combinations of the variables (e.g., $H = aT+bU$ where a and b are constants). The rules governing the form of useful functions and coefficients have yet to be formulated for sets or classes of compounds, but it is clear that such rules will be established by curve fitting from the existing world data bank on intrinsic toxicities and exposure modalities. (Exploration of this subject should be reserved for a separate study.) It will be of prime interest to discover the degree to which the actual interdependence patterns of T and U factors for specific compounds reveal which of the two independent variables predominates in contributing to the absolute magnitude of the hazard H.

TYPES OF HAZARDS

The earlier definition of hazard indicates that it is a function of the toxicity of a chemical and its use. In the statistical sense an expression of hazard would be a continuous variable. However, for the first cut regulatory purposes, it is useful to categorize hazards into classes and ignore their continuous nature. The three useful classes are: high-likelihood, low-likelihood, and uncertain hazards.

1. *High-likelihood hazards* are those where the dose rate exposure effects are generally predictable. Observation of exposed populations will generally permit identification of a threshold dose, and exposure above that threshold dose is likely to result in a large fraction of the exposed population manifesting the effects. Examples of such hazards are the cholinesterase inhibitory effect of organophosphate pesticide exposure, the narcotic effect of vinyl chloride monomer exposure, or the encephalopathy effect from lead exposure.

2. *Low-likelihood hazards,* those of primary interest in this report, are much less predictable. Observation of exposed populations is much less likely to result in the identification of an effective threshold dose. The dose rate and exposure effects tend to become manifest in the longer term and to be irreversible once manifest. Examples of such hazards are the purported chronic central nervous system effects of the organophosphates, hemangiosarcoma of the liver after vinyl chloride exposure, and possibly the carcinogenic effect of DDT exposure.

3. Exposure effects in the *uncertain hazard* category are poorly predictable in that they are second- and third-order effects or represent interactions with other chemicals. Examples of such uncertain hazards are the effects of freons on the ozone layer; the combined effects that result from concomitant exposure to sulfur dioxide and suspended particulate matter, including the possible additional tumorigenic effect when that particulate matter contains benzo[a]pyrene as well as other trace metals such as iron; and the synergistic effects of increasing the frequency of lung and gastrointestinal tract cancer among persons who have had asbestos fiber inhalation exposure and who smoke cigarettes.

IMPLICATIONS OF THE HAZARD FUNCTION CONCEPT FOR CONTROL

Given the ability to generate a credible value of H for a given chemical exposure modality, it will be of great regulatory significance to assess the manner in which the imposition of a control measure C on permissible concentration levels causes change in the magnitude of H. In a sense, the control C has to be viewed as an enforced modulator of the hazard H, and it is imperative that the sensitivity of H to change as a function of increments in C be ascertained. This sensitivity relationship will be a powerful tool in the hands of the regulator. Carried a step further, if H does vary sensitively with C, it may well turn out that very useful functional relationships of the form $H = f[T(C),U(C)]$ may be constructed and used to interrelate all useful parameters in quick assessment of hazard to man.

It should also be noted that the control function C operates on toxicity T only when a mixture of chemicals, as in a product, is being considered, and not when the intrinsic toxicity of a pure chemical is being evaluated.

Referring now to the three degrees or categories of hazard posed by toxic chemicals in man's environment, it is important to note that the concept of separate toxicity and usage contributions to a hazard rating index has important operational consequences for the regulator/decision maker. First, for the high likelihood hazard, the T factor will be high and so will the H index. Therefore, it becomes important to control the level of hazard H by stringent control over the usage pattern U.

For the hazard category designated as the low likelihood hazard, the reliability estimates for both the T and U data sets are of critical importance in setting up control strategies. Any exposure may yield some evidence of toxicity. It is therefore profitable to diminish or prevent hazard in this category by control limits on permissible exposure, since the number of projected cases of toxicity encountered will be a function of a probability factor operating on the size of the population segment exposed. However, it should also be noted for this category of chemical or product that an unusually high benefit attribute may well generate a different societal attitude toward minimizing hazard by limiting usage and exposure. With widespread use of such a chemical with high benefit indexes, it might well happen that a "socially acceptable risk" concept could evolve in a considerable sector of the population.

With regard to the third category of chemicals, characterized by hazard indexes that are uncertain or poorly predictable, special problems arise for the regulator. Basically, the amount of testing required to establish reliable hazard ratings under all exposure conditions of importance may be too extensive and expensive to be accomplished within available resources, except for a very few chemicals with special priority tags. This situation is best resolved by a recommendation for continued surveillance of potentially exposed populations for the incidence of toxic effects that might be elicited by either new or existing chemicals.

FACTORS INVOLVED IN HAZARD ASSESSMENT

The Data Base

The size and nature of the data base required for hazard assessment will vary with the type of chemical, the availability of pertinent information (toxicity, persistence, and so on) on known chemically related materials, and on the current and/or projected uses.

In establishing the toxicity characteristics of a chemical, a sequential series of testing steps could be taken (NAS 1975). Results of preliminary studies are likely to indicate the direction to be followed by subsequent studies. In those cases in which the producer desires to accelerate the program, testing steps can be taken concurrently. This approach may result in undertaking some studies which sequential testing would have indicated was not required, but the additional information will be desirable.

In assessing the hazard of a chemical in current use, effects on human health can be predicted from results of tests using traditionally accepted test animals to establish acute and chronic exposure effects; the effects can be confirmed by appropriate studies of people who have been exposed to the chemical. The exposed population should include the individuals involved in the manufacture, handling, and use of the chemicals, the consumer of the ultimate product, or the general public. The clinical and epidemiological data derived from these studies are also particularly useful in estimating the economic losses relating to human health considerations. Human health assessment of a *new* chemical must be based on similarities to known chemicals, on results obtained from studies involving test animals, and on projected use.

Environmental hazard assessment must be made using information relating to effects observed on ecosystems or the biota and confirmed or predicted by results obtained from appropriate laboratory tests (NAS 1975). Except for endangered species, environmental hazard assessment for the biota should focus on maintaining adequate populations of each species rather than on adverse effects on individuals. To avoid secondary effects, the natural balance between trophic levels must be maintained. Furthermore, mutagenic effects must be seriously considered, whereas teratogenicity and oncogenicity are probably less important. For those chemicals that are stable and that therefore could persist in the environment, the impact of bioaccumulation and biomagnification must be established since presence per se cannot be assumed to be a hazard.

Other effects which do not involve toxicity considerations but which are nevertheless undesirable are those relating to damage to materials

(NAS 1975). Examples of such damage would include the fouling of recreational waters and beaches and the discoloration and deterioration of man-made structures. Assessment of this type of loss must be included along with hazard assessment involving toxicity and use.

Reliability of the Data Base In establishing both the toxicity data base and the use data base, the reliability of the available information must be critically evaluated. Laboratory test data will always reflect the limitations common to such studies. Extrapolations of results observed with one species of test animal to another species are often tenuous because of species variations to chemical exposures. Studies of human exposures, unless carefully designed, are of necessity based on incomplete information, much of which has been accumulated for other purposes. The complex competitive nature of the marketplace inhibits the efficient gathering of information for the use data base. For widely used and persistent chemicals, the generation of material balance analyses can often be useful.

Developing the Data Base The preparation of the data base must be shared jointly between the producers of the chemical and the appropriate government agency. For a new chemical, the producer will be the logical source for most of the information. For a chemical which has been used extensively over a period of time and distributed by many producers, the government agency is likely to have the resources required to compile the data base.

The competitive nature of the free enterprise system prevents the open exchange of product information between producers and government agencies. Equitable means of divulging proprietary information must be devised. These could include the acceptance by the government agency of information from the producer on a confidential basis, the payment—from competitor to original producer—of a fair share of the cost of obtaining the information, or the joint funding by all producers of any information-seeking program. In establishing the use data base, difficulties will be encountered in determining the amount of chemical delivered by a producer to each of his customers, since these data are considered to be marketing information vital to the survival of the producer's business.

New Toxicological Procedures

The development of new concepts in toxicological procedures must be continually encouraged. These developments could originate in any qualified laboratory, but the federal regulatory and research groups must share the leadership in encouraging innovation and establishing programs to evaluate these new developments and to further their evolution into valid techniques for generating additional information that will lead to registration. Upon scientific acceptance, these new procedures must be immediately included in the information-seeking process. The federal government must ensure that adequate resources are available for the development of new concepts. Research agencies must be sensitive to the needs of the regulatory agencies. The application of new toxicological procedures to an existing chemical will require judgments relating to the appropriateness of the new procedure to the chemical and its uses. If deemed appropriate, the governmental regulator must assume leadership in initiating the studies, using its resources or, preferably, arranging for a program in which the producers can also participate.

Safety Margin

Safety margin refers to the incremental stringency factored into a permissible concentration of a toxic substance to compensate for uncertainty. The concept of a safety margin is most useful when the toxic hazard fits into the previously discussed high-likelihood hazard class, but it may also have some applicability when it fits into the low-likelihood hazard class with specific respect to exposure. The uncertainty may result from the ascertainment of the threshold concentration for toxicity. For example, if the threshold effects data were gathered on one or more animal species, a larger safety factor would be necessary in using these data to assure human protection than would be necessary if the available data had been gathered from human exposure situations. Another kind of uncertainty that must be considered and compensated for in choosing the magnitude of a margin of safety is that which arises from speculations as to the use (including future uses) of the substance and the possible exposure concentrations that could result therefrom.

The magnitude of a safety factor has associated cost considerations. The cost trade-offs that must be considered are those associated with assembling more precise toxicological and use data versus the incremental costs that will be incurred by requiring more stringent controls because of a less precise data base. On the average it will cost more to institute the more stringent control than to improve the precision of the data base. This option is removed when the operation is carried out under legislatively mandated short deadlines.

New Use Patterns

Changes in the use pattern of a chemical could result in the introduction of a hazard not previously experienced or predicted. Since this possibility exists, each new use must be treated as if the chemical were new, and a complete hazard reassessment must be made.

VALIDATION AND REFINEMENT OF HAZARD ASSESSMENT

To be successful, hazard analysis must have a sound background in both theory and fact. Insufficient attention has been paid in the past to the science involved with this process.

Theoretical Problems

The paramount theoretical uncertainties of hazard analysis lie in the area of low-likelihood hazards. They consist of problems of the effects on humans of small amounts of a chemical which is carcinogenic or mutagenic in laboratory animal studies. Do thresholds—concentrations below which no effects occur—exist for these compounds? If thresholds exist, regulatory decision making is facilitated, since regulation of use can control the chemical hazard. Scientific opinion is divided, but it seems to be approaching the conclusion that in a practical, operational sense thresholds for these effects do not exist.

Adequacy of Data Base

The broader issue concerns the accuracy and precision of the estimates of human toxicity based on data obtained in laboratory studies. Many scientists are increasingly confident that these studies predict more accurately and precisely for man than has been admitted in the past. The need is not for more debate but for more data. The scientific communities—in government, industry, and academia—must systematically compare the results of laboratory animal studies with those results seen in man following deliberate or inadvertent exposure. This requires a unique blend of toxicological, clinical, and epidemiological competence and cooperation which is unfortunately rare today.

Similar considerations apply to validation of estimates of "use" in its broadest sense. How precise and accurate were the estimates? Monitoring of the chemical and comparison of results with predictions of effects become of major importance.

If the reliability and precision of hazard evaluations, including toxicity and use, can be increased and understood, regulatory decision making will become, if not a simple task, at least one that man can aspire to perform with credit and confidence.

Because of the large number of chemicals that require evaluation and assessment for regulatory decision making, the regulator usually cannot wait for the full development of all information desired. Thus initial estimates of hazard and benefit tend to be less precise than desirable. However, even if this were not the case, ideally it is still highly desirable that an appropriate feedback monitoring or surveillance system be established that can: (1) reevaluate and reaffirm and thus validate the initial estimates; (2) assure that permissible concentrations which protect exposed and susceptible populations from effects are being maintained (these concentrations can be below a threshold and theoretically can be completely protective, or they can be an assessment that no more than an accepted amount of population risk is, in fact, being incurred); and (3) assume that the actual levels of exposure occurring in the community are not responsible for some new and unexpected effect. It should be emphasized that the only way to identify risks that have escaped detection by prescreening or other toxicological testing is by epidemiological or population surveillance studies.

Operationally, because of the large number of regulatory actions, much of this feedback will come from monitoring the ambient environment with acceptable methods in appropriate places; and only with even greater selectivity can human surveillance systems be established. This limitation makes evident the need to develop and use biological monitoring systems that can track and determine the time trends for concentrations of specific toxic substances in biological tissues and the need to develop and use population surveillance systems that reflect changes in illness and death patterns due to environmental pollutant exposure. Also it should be evident that such population studies provide a way to refine and update the performance of the cost analysis.

COST ANALYSIS

CONCEPT, MEASUREMENT, PRESENTATION, AND DISTRIBUTION OF COST

As a generic concept, cost is a diminution of something of value. In the context of regulating chemicals in the environment, a somewhat more specific construction of the concept is possible. For this purpose, cost should be construed as forgone human and environmental opportunities. In essence, the principle should be to count as a cost those reductions in valuable opportunities that can be traced to a change in the availability of a chemical in the environment.

Viewed in this light, there are costs associated with the decision to restrict or eliminate the use of a known substance and costs associated with the decision to permit its use. Costs cannot be avoided. Decision making can only affect the volume and distribution of costs.

Measurement of Costs

It also follows from this broad view of costs that not all costs are capable of being expressed in a common unit and that perhaps some types of costs cannot be measured at all. For example, a regulatory decision may facilitate or constrain the attainment of a social goal which is itself unquantifiable. (Later comments on the possible implications for the structure and concentration of industry are a case in point.) It is not surprising then if the effect of the decision is not quantifiable. A goal of cost analysis should therefore be to organize the potential effects of decisions using as few units of measurement as possible and to devise consistent ways of portraying those that are not quantifiable. A two-way classification scheme is possible that organizes the analysis under headings of pecuniary and nonpecuniary, provided it is understood that the latter group may be stated in several modes. This will avoid attempts to convert into a pecuniary measure those costs which are conceptually, philosophically, or technically difficult to force into such a mold.

Presentation of Costs

The form of presentation of the cost analysis influences the decision process. What is too much or too little or too varied a portrayal of costs is determined by the decision process. Therefore, the form of presentation is itself in part a policy decision. It should not be treated as a purely technical one and should be a conscious choice of decision makers. In the case of a new product, the costs of an affirmative or a negative decision are likely to be more conjectural than for a product already in use, but a common set of cost categories should be considered even though some may have to be assigned a zero value or contain only qualitative information or rank orderings.

Whose Costs?

The diverse nature of costs not only requires that they be expressed in different units, it also requires that it be clear which types of costs are relevant for which decisions. Conventionally, distinctions are drawn between those costs which are real, i.e., those that reduce the opportunity set with which society is faced, and those that are purely monetary, i.e., those that leave society's opportunities unaltered. This is a useful distinction for some decision processes but not for others. For example, it is a useful distinction in noting that the cost of health care to individuals is a real drain on society's ability to produce. And it is useful to discern the difference between the loss of income to an individual whose working life is interrupted or shortened and the fact that this may entail no loss to society if otherwise unemployed individuals are put to work, making social output the same in either case. From a social standpoint, the latter situation reflects a monetary transfer and not a real cost.

However, from the standpoint of individuals the reduction in income is a cost. The decision maker must therefore be certain which standard of cost assessment is involved. Clearly, in many instances, multiple viewpoints on cost are relevant and no conceptual error is involved in viewing an interpersonal transfer as a type of potential cost even when there is no direct real cost involved.

Similar distinctions have to be made in the use of market prices as measures of costs and adjustments that have been made in prices to reflect externalities that are not captured by the market mechanisms. Market prices that have been adjusted in this way are a more accurate reflection of true resource cost, but the unadjusted prices are more relevant for the analysis of the impact of a prospective action on individuals, enterprises, area income, or employment.

Stated generally, what is a cost is not independent of the vantage point adopted. The function of the decision maker is to weigh the impact on alternative groups, so it follows that even when the costs can all be stated in a common, pecuniary unit, they may not be additive. They may arise from different initial suppositions about the perspectives adopted.

It is therefore a useful principle to require that cost analyses be prepared that clearly identify the vantage point adopted, and that for each potential decision, a collection of relevant vantage points be initially adopted as a policy decision. For example, the analysis of hazard will necessarily have identified the groups for whom a potential toxic effect exists. Beginning there, an understanding of the parts of the economic, environmental, and social systems with which the directly affected groups are importantly involved can identify other perspectives from which costs may need to be analyzed. The fundamental point is that what constitutes a cost depends upon whose valuable opportunities are reduced, and that the units for measuring that reduction were dictated, in part, by the perspective adopted. Great caution must be exercised because costs that are commensurable may not be additive, and in any given analysis, commensurable, incommensurable, and intangible costs may all be relevant. To deny this is equivalent to deciding to reach a decision on the basis of incomplete information.

TYPES OF COSTS

The foregoing explains the difficulty of specifying the appropriate range of costs which should be considered in any given case. In place of this, it may be more useful to provide a categorization of types of costs which may be encountered. For those who wish to do so, this might serve as a checklist of things to consider in making the professional judgment of how wide a cost-catching net to throw in the process of a particular analysis. Though other categorizations might prove equally useful, the following four-way system is one alternative.

Direct Noneconomic Costs

The physical effects on human life, human health, ecosystems, and the materials of the toxic substances at issue will have been identified by the hazard analysis that in turn triggered a cost analysis. By the nature of the types of costs involved, most cannot be usefully expressed in monetary units; but they should be described (and where possible quantified) in units that convey their essential noneconomic character. Where there are unique susceptibilities (for example, among infants or the aged), hazard analysis will have made them apparent. This determines the subcategories of cost to be considered. In transferring these data to a cost analysis, however, particular attention should be paid to the comparative effects of alternative control devices or of substitute products that may be induced into use. For example, it may not follow that elimination of a hazard will lead to a total reduction of cost if the result is a substitution that produces costs either within this category or in any of the four cost categories.

Direct Economic Costs

These may be conceived of as falling into two general types, transactional costs and control costs. The former are probably the easiest to estimate, because an adequate first approximation involves the direct regulative and administrative costs of toxic substance control. These are mandated by public law, and some are initially incurred prior to the determination of whether regulation is even necessary. Whether or not additional transactions costs exist depends on the regulatory option chosen;if they do exist, their magnitude may itself be a factor in selecting the appropriate regulatory action.

Control, or what might also be termed compliance, costs consist of a more complex set of direct cost effects. In general, these are the impacts related to those directly involved in the production and distribution of the substance at issue. These can frequently be most easily perceived as changes in output, but even here the nature of the production process involved requires examination of the possible impact on jointly produced products and on by-products, including wastes. Closely related to these changes in the value, volume, and durability of output (in fact so closely that care must be exercised to protect against double counting) are income and employment changes whose magnitudes can be used in some analyses as surrogates for each other and for output changes themselves. However, as indicated below, there are distributional aspects of these measures which are not as readily perceived if only output changes are captured by the analytical framework.

From the nature of control costs it is obvious that they can be minimized the earlier the decision to regulate is made in the sequence between innovation and introduction. This is true of both the direct economic costs and the indirect economic costs which are discussed below.

Any regulatory decision short of outright banning of a toxic substance implies that continued use is likely to induce some direct costs for health care for those adversely affected. This is as much a direct cost of adjustment to the hazard as are monitoring costs or output changes, and it should be reported as such.

Indirect Economic Costs

These costs are considerably more difficult to assess, and attempts to include them in the analytical framework should be preceded by a judgment that the apparent balance between costs and benefits is likely to be altered by their inclusion. In general, these indirect economic costs arise from the commonly observed phenomenon of linkage among economic units.

It may be useful to visualize these linkages as arising first from the fact that products are frequently either complements or substitutes for one another. In the case of complements, use of one substance makes likely the use of another, so that regulation of one may have notable effects on the use of the other. Alternatively, when substitution is feasible, regulation of one may radically alter the use of the other. In the latter case, there may be hazards induced through a change in use that may require a complete cost analysis.

Another linkage is that which exists because firms or industries are customers or suppliers of each other. The regulation of a product conceivably might have a negligible direct impact, but it may be a critical input to other producers. In other cases, those that formerly supplied inputs for the controlled substance may experience such increases in unit costs as a result of decreased sales that the cost and availability of other products using this same input may be importantly affected.

These effects, which are a route through which regulatory impacts ramify through the economy, are also captured by changes in trade relations. The volume and direction of trade internally and with foreign countries can obviously be altered if existing buyer–seller relationships are altered by regulatory decisions. To fully assess costs that may be associated with these decisions, it may be necessary to estimate not only the adjustments in production inputs and outputs that could result but also what those adjustments imply for trade among internal and external geographic areas. There is therefore a regulatory need for quite detailed information about trade in toxic substances for use in both hazard and cost assessment.

This same trade information could be highly useful should it be judged necessary to inquire into another variety of indirect economic costs—those of a distributional nature. Implicit and explicit in what has been said is that regulation can affect the relative fortunes of individuals and of areas. Even when there may be no overall social costs of a regulatory action, there may well be distributional costs to some and gains to others. Their distribution among areas and economic and social classes may be important data for regulatory decisions even if comparable distributional data on benefits may not be available.

Regulatory decisions can also have impact on future generations of humans. This is obvious when hazard assessment reveals mutagenesis as a likely effect, or when the fertility of populations is being adversely affected. Less obviously, intergenerational impacts may result from regulatory decisions that affect natural resource use and technological innovation and application. For that reason intergenerational effects are also involved in the next category of costs.

Structural Costs

Both directly and indirectly, the regulation of toxic substances can have impacts on the number of firms that compose the chemicals industry and on the extent to which the production of given products is concentrated or dispersed among firms. By law, for three-quarters of a century a goal of national policy has been the preservation of competition, and regulatory behavior can wittingly or unwittingly adversely affect this goal. A potential cost to be considered therefore is the likelihood of adverse competitive impacts from regulatory actions. The mechanisms that can translate regulatory controls into competitive effects are numerous, ranging from the inability of some firms to make the capital investment needed for hazard reduction to the possibility that a series of decisions will gradually deplete a firm's product line and leave it in an unsustainable position. There are two attributes to note about this possible competitive effect. First, the cost consequences that would be captured as direct or indirect economic costs are not the sole conceivable economic consequences of regulation. And second, regulatory decisions may have cumulative impacts that are difficult to associate with any single decision.

This latter point is at the heart of the concern that another variety of structural cost may be incurred. If the effect of regulation is to leave industry more concentrated and to deter new firms from entering the industry, entrepreneurial behavior will likely be reduced, and with it some of the dynamic qualities that we have traditionally expected as a social benefit from the pursuit of private profit. It is unlikely that any one regulatory decision could be associated with such an event, but cumulative causation could exist.

Another adverse effect would arise if the way in which regulation is conducted inhibited the search for new products and processes. The motive of private profit has been a social device for inducing invention and innovation. If regulation raises such difficult barriers that potential innovators are dissuaded from the attempt, social costs may be unwittingly incurred. Again, no one regulatory decision may yield costs of this type, but they may be produced by the form, philosophy, and mechanisms of regulation. The principle that suggests itself is that the administration of regulation itself may be viewed as potentially yielding indirect structural costs and that it should be analyzed from that vantage point.

ASSESSMENT OF COSTS

Availability of the required data is one of the most important requirements for cost analysis. Often, if not always, there is paucity of the data ideally required and frequently a complete absence of information over time.

In general, the data required for a cost analysis may be conveniently broken down into two major categories: (1) data generally available from well-established sources but requiring updating, and (2) data not generally available, except from such ad hoc sources as special studies.

This breakdown is, of course, arbitrary, and often data included in the first category may be available only on a limited basis. The production capacity information for an industrial sector is an example of this. Conversely, some of the data included in the second category may be readily available from some industrial associations. In general, however, data in the second category will require considerable analytical effort, while those in the former can be obtained from the standard governmental publications.

Data Likely To Be Required

1. Data generally available requiring regular updating: output trends; employment trends; wages, salaries, and earnings; interindustry impacts (buyers and sellers); structure of industry; industry (and firms) product line and diversification; exports; imports; regional aspects; and industry capacity.
2. Data possibly required and/or special inquiries needed: elasticity of demand; elasticity of raw material supply; "age" of technology employed; age of capital; use of plant for other products; financial characteristics of firm and industry; costs of illness—real and financial; sociodemographic characteristics of affected populations; transactions costs; control costs; end-of-the-line controls; process modification/technology change; product change; and raw materials and input change.

As already indicated, data in the second category are not always readily available. Indeed, to determine the information listed in these 10 categories it is almost always necessary to perform some analysis, often using incomplete information. Particularly difficult to determine are the data related to "costs of illness—real and financial." Here the difficulties stem not only from a paucity of valid empirical information, but also from conceptual problems. These problems are aggravated by the fact that this information is of paramount importance to the analysis. Similar conceptual problems exist for the data listed in "control costs." The conceptual and related problems of controlling hazardous environmental chemicals are of sufficient importance to merit further discussion.

There are a number of alternative hazardous chemical controls, each of which is associated with unique control costs possessing distinct advantages as well as shortcomings. The most frequently used approaches to control of hazardous chemicals are as follows:

Treatment This is the most common method used to control chemical residues as well as common pollutants. It has one basic disadvantage in that removal of some chemicals may require expensive treatment equipment. For chemicals in waste water, it is not likely that primary and secondary treatment will achieve the required reduction. The use of tertiary treatment may therefore be required. However, the high cost of this treatment may be an obstacle.

Recovery and Reuse This method is a variant of the treatment approach, and it requires in addition the recovery and reuse of chemicals.

Product Modification This method is defined as a "deliberate introduction of new properties into produced materials to reduce their toxic effects or to enhance their controllability." The method involves adding or removing chemicals or substituting alternative materials. It would not be applicable where alternative materials did not have the required chemical or physical properties.

Changing the Production Process This method requires modification of the production process so that a chemical is not given off or is emitted in reduced quantities.

Elimination This method calls for outright elimination of the use of chemicals and therefore is closely related to the product modification method discussed above.

Other Control Methods In addition to the five methods described above, six others are available for abatement of most pollutants but are not readily applicable to most toxic chemicals. These are (1) dispersion, the distribution of pollutants over a larger area; (2) dilution of the volume of pollutants; (3) detention, the temporary holdup or storage of pollutants for later gradual release or release at a more advantageous time; (4) diversion, the transportation of pollutants to another location for treatment and/or discharge; (5) environmental treatment, the treatment of the environment to remove pollutants, diminish their effect, or eliminate or inhibit their generation; and (6) desensitization, rendering the potential pollutants harmless through desensitization of the receptors.

All these methods have been employed with success in abating conventional water and air pollutants. It is obvious, however, that most of them cannot be applied to toxic chemicals because of their inherent harmful characteristics.

The five methods most frequently used for controlling chemical pollutants must be evaluated through cost effectiveness analysis. Even a cursory analysis indicates that not all these methods can be used to control emissions from all sources. Instead, specific chemicals and specific uses call for different control methods.

The following six general factors appear to be the most important in evaluating the costs of alternative control methods: technical feasibility, direct costs, indirect costs, impact on the economy, quality of the end product, and time requirements.

Technical Feasibility Each control method must be suited to the technological conditions at the source of chemical pollution. These conditions include the availability of the required equipment and machinery, materials, research and development, and technical skills. Furthermore, these technological resources must be available during the control period as well as in the future.

Direct Costs Estimates of all direct and indirect costs determine the economic feasibility of any control method as opposed to its technical feasibility. Cost analyses must include both capital and operating costs. In estimating direct costs, two variables appear to be particularly significant: the level of technology (old, prevailing, advanced) employed by the industries which emit chemicals; and the size of these industries in terms of the quantity of pollution and the volume of the carrier of these emissions (waste water, air). Since both of these variables change over time, the direct cost estimates must also be analyzed in terms of the future.

Indirect Costs These are costs resulting from any control method which affects other sectors of the economy. The following two economic sectors may be affected more than others by indirect costs: suppliers of goods and services to the industries using chemicals, and buyers of the products of these industries. Analysis of the indirect costs involve factors similar to those considered for direct costs.

Impact on the Economy A cost analysis should include, where applicable, such national economic considerations as the impact of control methods on a country's gross national product, balance of trade, and other national economic programs.

Quality of the End Product Any control method may change the quality of the end product manufactured by several sectors of the economy. Two types of industries are particularly affected: (1) industries manufacturing goods that result in chemical residues, and (2) industries which use these end products as inputs for their own production.

Time Requirements In applying controls, time factors to be considered include: time to supply the equipment, machinery, operating supplies, and materials; time for construction and installation of the equipment; time to obtain technical skills; time to introduce a new product on the market; and time for the public to accept new products.

Finally, some of the required information, such as "financial characteristics of firm and industry" is regarded as confidential and is therefore difficult to obtain.

In light of these data problems and the cost of overcoming them, cost assessments should proceed with available data initially, but effective decision making will require investments in data acquisition.

RELIABILITY OF COST INFORMATION

Many costs are not subject to direct measurement, and some have been shown not to be quantifiable at all. Consequently, at best, cost data must be viewed as partially judgmental, and even where measurement is possible, it is usually imprecise. This is not as great a handicap as might first appear once it is recognized that regulation is not a mathematical procedure. Cost information can be good enough to provide the regulator with an understanding of the relative magnitudes involved and of the form and direction of probable adverse changes. For the information to be that good, however, requires conceptual skill and investment of resources to specify the data required and the costs and benefits of obtaining it.

Many of the data to be employed in cost analyses are economic in character, and there exists an enormous mine of such information. Its accuracy is probably inversely related to the specificity of the question asked. Much of it has been collected on a sample basis, so that its relevance to individual cases is diminished; and much of the rest has been intentionally masked to protect privacy rights and proprietary information. Interaction between technician and decision maker is an essential requirement to ascertain the extent to which generally available data will suffice and to determine precisely what additional data items are essential.

Collection of special-purpose data requires investment in skill and resources by both inquirers and respondents. It also taps a limited reservoir of good will. The data to be gathered, therefore, should be specified precisely, and the data-gathering process should include procedures for assessing the validity of the results. Varieties of devices have been used to cross-check on the accuracy of replies; and if these are not part of the data-gathering exercise, the reliability of the results may be undermined.

The usefulness of data may be assessed in terms of their accuracy, but they may pass that test and still not be relevant. Relevance is a function of the questions to be asked. Unless the regulatory questions are specified in advance, accurate but irrelevant data may result. An important attribute of relevance is the level of specificity or aggregation at which the data are collected. As a simple example, the mean of distribution may be adequate for some questions, but the variance around it may be critical for others. Similarly, it may be necessary to have highly disaggregated data about product flow within the chemicals industry and between it and its immediate suppliers and customers. Beyond that, however, increasingly aggregative data may be sufficient for a variety of reasons, an important one being the pertinence of third- and fourth-order effects to the decision to be made.

A discouraging attribute of economic data is that it is typically old, i.e., not representative of the immediate situation, by the time it is processed into usable form. And its usefulness declines rapidly thereafter. Historical data series are important, of course, but for regulatory decision making the unattainable ideal would be to know what is happening now to better gauge what may happen tomorrow. This suggests that continued investment in data to assure the optimum usable level of timeliness must be part of the transactions costs of toxic substance regulation. At best, the results will be imprecise, and the cost analyses that result should be considered as probabilistic estimates of the magnitudes under examination.

VALIDITY OF COST ESTIMATES

A number of factors may affect the validity of cost estimates. In general, they relate to an inability to trace cause–effect relationships. Two varieties of this problem are particularly troublesome to those who attempt to assess the effects of regulations once imposed.

The first arises because the economic system is dynamic, and errors can frequently be made if causal connection is inferred from chronological sequence. Hence, we cannot be certain that dynamic changes were initiated by toxic chemical control efforts simply because they followed those efforts in time.

Second, when firms alter their production facilities, it is frequently unclear even to them which part of the cost of change is related to what motivation. A good example of this is the costs of a new plant employing a new production technology that reduces production costs *and* limits the emissions of certain toxic chemicals. Another example is in the case of costs associated with recovery of by-products from wastes.

In the first example cited above it is not at all clear whether such costs can be attributed to the environmental control or to the profit motive of the firm. The result is clearly a "joint product" of the two—and a difficult concept for economic analysis.

In the second example, the firm undertaking by-products recovery may or may not report the associated costs separately. More likely than not, such expenditures are not broken out from the general capital account.

As a partial alleviation of these problems, regulatory agencies should consider the extent to which engineering cost estimates and simulation can provide independent estimates of costs that are otherwise difficult or impossible to obtain.

REFERENCE

National Academy of Sciences (1975) Principles for Evaluating Chemicals in the Environment. Washington, D.C.: National Academy of Sciences.

APPENDIX

H Working Paper on Hazard–Cost–Benefit Comparisons

INTRODUCTION

Decision makers such as the Administrator of the U.S. Environmental Protection Agency (EPA) are regularly faced with the necessity of making choices involving issuing regulations, setting standards, and so forth. These choice problems have three essential characteristics: the necessity for making (or accepting) trade-offs; the noncommensurability of the effects being weighed in the trade-off; and uncertainty or lack of information about the consequences of alternative courses of action. Whenever there are choices to be made, decision makers must somehow cope with the problems posed by these three characteristics of the choice problem.

This panel has been asked whether there is some kind of framework or quantitative language which can assist decision makers in carrying out their responsibilities. We believe that the systematic application of the tools of decision analysis and benefit–cost analysis can provide the decision maker with a useful framework and language for describing and discussing trade-offs, noncommensurability, and uncertainty. This framework should help to clarify the existence of alternatives, decision points, gaps in information, and value judgments concerning trade-offs. Furthermore, this framework and its associated language should facilitate communication between the decision maker and his staff of analysts, and between the decision maker and the public.

Because terms such as benefit–cost analysis are used in a variety of contexts and with different meanings, it will be helpful for us to be specific as to how we define them. Following the terminology adopted by the Committee on Public Engineering Policy, we use *benefit–cost analysis* to refer to "an evaluation of all the benefits and the costs of proposed action" (NAE 1972:3). *Benefit–risk* refers to "that category of benefit–cost in which risks to life and health are an important [but not the only] component of costs" (NAE 1972:4). This definition of benefit–cost analysis is a much broader concept than the traditional economic accounting of monetary values in that it encompasses all possible positive and negative effects of a proposed action and is not limited to those which can be measured and valued in dollar terms. The traditional version of benefit–cost analysis as used by economists takes as given the single objective of maximizing the efficiency of resource utilization in producing goods and services for individuals; and it defines and measures all benefits and costs in terms of this objective. But there can be other objectives besides economic efficiency. And our use of the term benefit–cost reflects acceptance of this multiplicity of objectives.

Also we wish to make a distinction between decision theory and decision analysis. *Decision theory* is a set of analytical tools, based on utility theory and the inductive use of probability theory, which provides a rational framework for choosing between alternative courses of action when the consequences resulting from this choice are imperfectly known (North 1968). *Decision analysis* refers to the systematic analysis and evaluation of alternative courses of action drawing upon the analytical tools and insights provided by decision theory to deal with problems of noncommensurability and uncertainty. Thus, decision theory refers to a specific body of prescriptive theory and analytical tools derived from it, while decision analysis refers to an approach to benefit–cost analysis as broadly defined above.

We must make it clear that benefit–cost and decision analysis as we use these terms do not provide a rule or formula which would make the decision or predetermine the choice for the decision maker. Rather, we view benefit–cost and decision analysis as providing a framework and set of procedures which will help to organize the available information, display trade-offs, and point out uncertainties. In this way, benefit–cost analysis can be a valuable aid to the decision maker; but it does not dictate choices, nor does it replace the ultimate authority and responsibility of the decision maker.

Benefit–cost analysis could be used as an acceptable rule or formula for making choices only if the following conditions were met:

1. there was unanimous agreement as to the definition and measurement of welfare;

2. the trade-offs that characterize each choice could be described in terms of a single, objectively determined, cardinal measure of welfare such as dollars; and

3. there was perfect certainty about all consequences and all information.

The various welfare effects of different alternatives could then be added together, and the alternative with the highest welfare number would be chosen. There would be no room for disagreement over the choice, given the assumptions, because of the prior agreement on value and the perfect knowledge of consequences. If alternative A meant more eggs and less sausage, the agreed-upon values for eggs and sausage would be used to express this trade-off in a common unit of measure. And the correct choice would be obvious.

In general, however, choices represent trade-offs among noncommensurable magnitudes (e.g., increased crop production versus decreased human health), and consequences are not known with certainty. We turn now to a brief discussion of the ways in which values for noncommensurable effects and uncertainty enter into rational, systematic analysis of choice.

VALUES AND CHOICE

Consider a hypothetical and highly simplified case of a pesticide in current use. Assume that the following information is known with certainty. At present use levels, the pesticide increases the net agricultural productivity of the land on which it is applied by $500,000 per year; but it causes 10 deaths per year in the exposed population. If usage were reduced 50 percent by regulation, productivity would fall by $200,000 per year and 5 deaths per year would be avoided.

This information can be displayed as follows:

Level of Use	Benefits of Use (Productivity, $/yr)	Costs of Use (Deaths/yr)
current level	500,000	10
50 percent control	300,000	5
ban or zero use	0	0

The problem is clearly one of trade-offs between productivity and human life. But in the absence of some agreed-upon basis for making deaths and productivity in dollars commensurable, no simple decision rule can be applied to determine the "correct choice." No objectively determined weights or values have been provided for converting these noncommensurables into a common unit of measure.

One way to approach the decision problem is simply to present this information to the administrator responsible for making the regulatory decision, and to allow him to make the choice on whatever basis he wishes. Suppose he chooses the 50 percent control option. This choice reveals something about the value system of the decision maker. It says that he is willing to give up $200,000 in productivity in order to save 5 lives, i.e., lives saved are worth at least $40,000 each in the decision maker's scheme of things. But because the decision maker chose not to go to a total ban, this reveals that he was not willing to give up an additional $300,000 to save 5 more lives. Thus his implicit valuation of a human life is less than $60,000. Because he had a chance to save 5 additional lives at a cost of $60,000 each and chose not to, he reveals what is to him an upper limit on the implicit value of human life. Different values for life would be implied by different choices. For example, the choice of a total ban would imply a value for life of at least $60,000.

In the example presented here, choice determined value, rather than value determining choice. The problem of valuation cannot be avoided. Our thesis is that since valuation of noncommensurables is unavoidable, it is better to confront the choice of values openly and explicitly than to allow values to be hidden.

An alternative procedure would be to present the decision maker with a display of information like that above, and to say: "The key variable is the value of human life. If you believe a life is worth something less than $40,000, you should not regulate use since the costs of regulation would be greater than the benefits," and so forth. Or if the decision maker says that he is leaning toward a total ban, he can see what this implies for value and test his intuitive preference for the ban against his notions of value.

This explicit examination of values may also lead to a more consistent pattern of choices over time as decision makers seek to rationalize the pattern of revealed values. Also the examinations of values may show some choices to be internally inconsistent. In the above example, suppose that productivity benefits of present use were $700,000 per year. A decision to impose 50 percent regulation implies a value per life of at least $80,000 ($700,000–$300,000 for 5 lives). Given this implied value, it would be inconsistent not to move to a total ban since the additional 5 lives saved would cost only $60,000 apiece.

It should be noted that relative values cannot always be uniquely determined from the available information.[1] Nevertheless, this simplified example illustrates an important point. Choices about trade-offs unavoidably involve placing relative values on noncommensurable effects such as dollars versus life. These values can be hidden if the information is not displayed in the appropriate framework, or they can be brought out in the open. The use of the proper framework for presenting information on noncommensurate effects can make the value implications of alternative choices clearer and allow the decision maker to confront the valuation problem openly and explicitly.

UNCERTAINTY AND CHOICE

A characteristic of most decisions to regulate toxic substances is that the consequences of the decision are not known with certainty at the time the decision is made. If chemical X is disseminated into the environment, will it be concentrated in food chains and ingested by humans? Will ingestion lead to severe consequences for human health? Will other species be harmed? Might the chemical cause physical changes in the environment, such as alteration of climate? When other chemicals are present, are the effects interactive? If the chemical is banned, will effective and economical substitutes be available to the users who have come to depend on it? What is the magnitude of the losses that will be incurred by firms producing the chemical? What will be the effects on employment, U.S. balance of trade, or research and development policies of producers? The usual situation is that regulatory choices must be made without definitive answers to these and other questions.

Benefit–cost analysis can be extended to deal with uncertain situations by the use of probabilitics to quantify uncertainty. This extension is straightforward in theory and intuitively appealing, because probability is a natural language for describing uncertain situations. People are accustomed to hearing probabilities of rain in predictions of tomorrow's weather or odds on forthcoming elections and sporting events. The same concepts may be used to describe uncertainties related to toxic substances. If a pesticide is registered for unrestricted agricultural use, the effect on human health is uncertain. An example of a probability statement to describe the uncertainty would be: "A 5 percent probability

[1]Relative values can be determined between any pair of noncommensurate categories of effects even when the total number of categories is greater than two, except when a pair of categories is available only in fixed proportions. The ability to infer relative values from observed choices is limited primarily by the availability of information on the range of alternatives. In the example above, the range of implied values could have been narrowed had there been productivity and mortality data on other levels of control.

is assigned that the compound is strongly carcinogenic in man and that additional deaths from cancer at the rate of 100 deaths per 100,000 exposed population would result from each year of unrestricted use; a probability of 5 percent is assigned that the chemical is weakly carcinogenic and would result in additional deaths from cancer at a rate of 1–100 deaths per 100,000 exposed population; and there is a 90 percent probability that the chemical will produce no increase in the incidence of cancer." In many situations the basis for such a statement may be subjective judgment based on information on total amount released, mechanisms for bioaccumulation, and a tentative dose–response curve for humans extrapolated from limited animal test data. The statement may be viewed as a summary of the available information on the toxicological consequences of the pesticide, and its probabilistic language expresses the uncertainty about possible effects.

The use of probabilities does not alleviate the difficulty in making major decisions on the regulation of toxic substances on the basis of scanty information. While this situation is acutely uncomfortable for the decision maker, there may be no way to avoid the issue. Inaction is a decision not to act. The necessity for decision forces the decision maker to confront trade-offs and uncertainty either implicitly or explicitly.

A FRAMEWORK FOR DECISION ANALYSIS

The main contribution of the decision framework described here is to organize information for the decision maker to assist him in this unavoidable balancing task. The framework requires that the analyst quantify degrees of uncertainty through the assignment of probabilities to various levels of hazard, and other possibly uncertain beneficial and negative effects, and then asks the decision maker to consider explicitly valuations of the health and other noncommensurable effects resulting from various levels of hazard relative to the expected economic benefits. This approach permits information from various specialists to be introduced unambiguously into the decision process: the probability of hazard from scientific specialists, the economic effects from economists and industry experts, and the trade-off between economic benefits and health effects by the decision maker responsible. The alternative is an implicit valuation, in which either the decision maker must assimilate the complex information relating to hazard and economic loss, or his advisors must impose their judgment on the trade-off between human health and economic loss.

To achieve a consistent approach to decisions on toxic substances, it seems preferable to have a framework in which the scientific, economic, and valuation assumptions are made explicit. An explicit framework permits the decision process to be reviewed by the various parties concerned, and where disagreements exist on the preferred decision alternative, the framework will help to reveal the basis of this disagreement: is the issue one of scientific information (e.g., the probability that the chemical is carcinogenic), or one of the valuation of cancer incidence versus economic benefits? Furthermore, as new information is obtained, the decision can be updated rapidly and efficiently by incorporating this new information into an existing analytical framework.

OUTLINE OF PRINCIPLES

In order to be of assistance to a decision maker, the benefit–cost framework and associated procedures should meet certain objectives and include certain characteristics. These objectives and characteristics relate to how the framework and procedures can be most useful to the decision maker in identifying information needs, organizing information, and making comparisons among alternatives.

1. The decision analysis should be based upon a simplified model, picture, or flow diagram of the total system of production, distribution, use, and disposal of the chemical. The model should help to identify points of economic impact, nature and source of benefits and damages, and possible means of control. This information will provide the basis for specifying alternative control strategies, and the quantification of costs, hazards, and benefits.

2. The decision framework should make it possible to identify and present information on the full range of alternatives the decision maker has, i.e., no action, partial control, or total ban. Alternative control levels as well as alternative implementation procedures and schedules should be included. The framework should display the data so that all relevant alternatives can be considered together. Other major factors that might influence the decision maker's choice, e.g., legal constraints, previous action, or ease of implementation, should also be identified.

3. The framework should include all identifiable effects and consequences of alternative actions. This would include social and economic benefits of the chemical's use, health effects, ecological effects, costs of control, economic impacts (plant closings, unemployment, Gross National Product), enforcement and monitoring costs, and distributional effects (who pays, who benefits). Care must be taken not to overlook relevant categories of effects. Use of a chemical may entail health

benefits as well as risks; and the benefits and risks might impinge upon different population groups. For example, an insecticide might control infectious disease vectors while having long-term carcinogenic effects after bioaccumulation. The decision maker may also wish to distinguish between risks borne voluntarily with full knowledge, and those borne involuntarily or without knowledge.

The level of detail to be included in the description of effects is partly determined by the decision maker to the extent that he determines the time and resources available for study and analysis, and partly by the quality, amount, and availability of data. For some purposes, only a brief list of effects need be considered; in others a comprehensive list would be necessary. The level of detail will also vary with the stage of the decision-making procedure in that a brief and quick analysis may be made to screen potential chemicals and then a more elaborate analysis conducted on those chemicals selected for more careful study. The framework should be flexible enough to meet these various needs of the decision maker.

4. The framework should facilitate the comparison of major effects resulting from the alternative actions and should serve as a convenient basis of discussion and review. To this end, the results of the analysis may need to be presented in different formats (e.g., discussion, graphs, tables) and at different levels of detail. The final briefing document may only present major effects and major alternatives. However, the most detailed analysis available should be provided as background material so that the decision maker can examine these details if he wishes to.

5. The framework and procedures should be flexible enough to meet the demands of different kinds of decisions.

6. All effects should be quantified and measured in commensurate terms to the greatest extent possible. Further, the number of noncommensurable measures should be as small as possible to simplify the trade-off considerations of the decision maker.

7. The framework should make it convenient to determine the value of obtaining further information and specifically what information should be obtained. In other words, it should make it clear that resolving uncertainty (collecting more information) is a relevant choice for the decision maker although it may involve some time and costs. In this way the framework will facilitate the process of continuous review or sequential decision making.

8. The framework should indicate the range of uncertainty and level of ignorance about key pieces of information. If detailed analyses are aggregated or summarized for presentation to the decision maker, information about the degree of uncertainty must be presented.

9. All assumptions made by the analyst in reducing detailed data to summary or aggregate measures should be made explicitly and clearly indicated as such.

10. Where uncertainty exists in some key pieces of information, or where assumptions must be made about the relative importance of certain effects, the decision maker should be able to examine the sensitivity of the results to variations in both input data and different assumptions about relationships.

11. The key value judgments about weights or values to be assigned to noncommensurables should be the responsiblity of the decision maker. The presentation of information should make it clear what the key trade-offs are and facilitate the examination of alternative value judgments by the decision maker.

VALUATION

VALUES AND VALUE JUDGMENTS

A choice problem where there are two alternatives, A and B, can be described by a list of the differences between what is expected to occur if one or the other alternative is chosen. Alternative A may be expected to yield higher agricultural productivity than B, and a larger number of cancer deaths. Filling in this list is basically a problem of quantification and measurement.

Some of the effects may be viewed as beneficial and others as adverse. A decision to choose A implies that, in comparison with B, the beneficial effects outweigh, in some sense, the adverse effects. This judgment involves valuation as distinct from measurement. It implies that beneficial and adverse effects were compared in terms of some common measure, e.g., welfare. The term valuation refers to the conversion of disparate, noncommensurable effects, both beneficial and adverse, into some common measure through the use of weights or values. This section is concerned with the choice of the weights or values to be used in this process of valuation.

It must be recognized that the process of valuation is ultimately entirely subjective, at least for the kinds of decisions to be made by EPA and other government agencies in the area of chemicals and health hazards. For example, conventional benefit–cost analysis as applied to the evaluation of water resource investment projects provides a set of rules for calculating benefits and costs in dollar terms based on market prices. In this sense, the valuation of benefits and costs is objective. But the statement that conventional economic market measures matter for policy choice is fundamentally a subjective value judgment. Given that value judgment, objective rules can be used to assign values. But the statement that economic magnitudes matter for policy and, more importantly, the ways in which these magnitudes are compared to other kinds of benefits and hazards reflect basic value judgments.

SOCIAL WELFARE AND INDIVIDUAL WELFARE

One basic problem in the use of decision analysis or benefit–cost analysis is the relationship between collectively or politically determined values and individual values. At issue is the validity of social or collective measures of welfare as distinct from the welfares of the individuals in society. The western political tradition holds that individual welfare is what matters, and that individuals are their own best judges on this question. This position is the foundation of normative economic analysis (welfare economics) and conventional benefit–cost analysis. It follows that only those values which result from or reflect the voluntary actions of individuals, e.g., market prices, should be used in calculating welfare measures. But as should be clear from previous discussion by this and other panels, unambiguous individualistic measures of value are available for only a portion of the positive and negative effects of regulating chemicals. It is inevitable that decision makers must act as if there were some measure of collective or social welfare. And the values that are implicitly or explicitly part of their choices make up the social welfare function. These values are value judgments made by decision makers in a political context. It is the political process and the accountability of decision makers which must be relied upon to maintain some degree of correspondence between their collective value judgments and the values of individuals.

BASIS FOR ASSIGNING VALUES

This section discusses principles and procedures which can be used for assigning values to beneficial and adverse effects so that the remaining noncommensurate effects which are presented to the decision maker can be as few as possible.

A major class of consequences is changes in the availability of goods and services for individuals. These are termed economic efficiency benefits (where availability is increased) and costs (where availability is decreased). We assume that the total availability of goods and services does matter to the decision makers, i.e., it is a part of social welfare, and the value judgment is made that individuals are their own best judges of the value of additional goods and services. These value judgments make it possible to use rules derived from economic theory to reduce a substantial portion of the positive and negative effects of a regulatory decision to commensurate dollar values.

Market Values in Dollars

Under certain conditions market prices can be used as a means of assigning values to the goods and services produced. The assumptions are: (1) that markets are perfectly competitive, (2) that there are no external economies or diseconomies or other forms of market failure, and (3) that the distribution of income and purchasing power be judged optimal.[2]

The price an individual pays for a good indicates his relative marginal valuation of that good. Price then is a measure of private benefits at the margin. In an ideal world characterized by perfect competition in all markets, fully internalized costs and benefits, and the socially preferred state of income distribution, marginal private benefits are equal to marginal social benefits. Similarly, price is the additional revenue received by a competitive producer for selling an extra unit of output and the amount of revenue forgone by diverting resources to alternative uses. As long as the gain from selling additional units exceeds opportunity costs, a profit-maximizing producer will increase sales. Price then is a measure of private marginal cost, and in the absence of externalities and market failure private marginal costs are equal to marginal social costs. Thus, given the preferences of consumers and the technical possibilities of production, the system of prices which evolves in "perfect markets" will serve to maximize the value of production or economic efficiency.

These prices also measure the values of additional outputs (benefits) and values of opportunities foregone (costs). If decision makers are concerned with the efficiency with which the economy uses scarce resources for producing goods and services, then they would prefer those alternatives with the largest excess of money benefits over money costs, other things equal. Where the regulation of a chemical causes changes in market prices and quantities, this market information can be used according to the well-developed rules of conventional benefit–cost analyses to assign values to diverse effects, and to combine them into a

[2]For a more thorough discussion of the justification for using market prices as measures of value, see the papers in U.S. Congress (1969). Also, see Haveman and Weisbrod (1973).

single dollar measure of economic efficiency costs or benefits. For example, the market value of the labor and capital used in an emission control system can be used to assign a dollar value to control costs.

Shadow Prices

Where there are various kinds of market failures, there may be positive and negative effects on individual welfare which are not adequately reflected in market prices. Yet these effects should be included in a complete accounting of economic efficiency benefits and costs. Examples include air and water pollution and public goods such as national defense.

Because these nonmarketed effects are often linked with marketed goods through substitute or complement relationships, it may be possible to infer equivalent money values or shadow prices for these effects from information contained in market prices and quantities. This involves a kind of detective work—piecing together the clues about the values that individuals reveal as they respond to other economic signals. If appropriate techniques of measurement are used, the resulting values are fully commensurate with economic efficiency values derived directly from market prices.[3]

For example, differences in residential property values between areas with polluted air versus those with clean air may reflect the willingness of individuals to pay to avoid the perceived effects of air pollution at their residences. If these differences are derived from a statistical study that effectively controls for other factors influencing property values, they can be used to derive a partial measure of the cost of air pollution. It is a partial measure because it does not capture those effects which are so subtle and long term as not to be perceived by the individual, or effects perceived at places other than home, e.g., at work. Provided that double counting is avoided, these measures of efficiency benefits and costs can be combined with those derived directly from market price.

Economic Impacts

There are other types of economic impacts associated with regulatory decisions which can be described and quantified in dollar terms, but which are not commensurate with the economic efficiency benefits and costs described above. This lack of commensurability arises because the

[3]See Freeman (1973) for a description and evaluation of some of these techniques.

impact measures are usually based on changes in the dollar volume of a particular class of economic transaction rather than changes in the quantities of goods and services available to consumers. This distinction is similar to the one often made in the benefit–cost literature between real and pecuniary effects (Haveman and Weisbrod 1973).

For example, suppose a chemical plant closes because of regulatory action, and an annual million-dollar payroll is lost. This is surely an economic impact to the region; but it is not necessarily a real cost in the sense of a reduction in available goods and services. The relationship between the impact measure and the cost measure depends on the mobility of labor and other resources and the speed of adjustment. And these factors will vary widely from case to case. In the general case, where a firm or an industry reduces output because of regulation, there is an economic efficiency cost to the extent that resources remain unemployed or underemployed for some period of time. The forgone output or opportunity cost is the appropriate measure of cost. The cost of resources absorbed in relocation or other activities designed to speed the adjustment process (e.g., labor retraining programs) should be added. If all workers laid off from the plant immediately find alternative employment at the same wage (reflecting productivity) somewhere else in the economy, there is no real cost, other than the cost of relocation. If none of the labor can find alternative employment, then there is a cost in the form of lost output, and the lost payroll may be used to measure this cost.

It is possible that decision makers would wish to consider compensation to those resources which lose substantial income because of EPA rulings. A similar compensation principle has been part of U.S. international trade legislaton since 1962. However, while such compensation payments would represent a claim against government revenues and should be displayed for the decision maker, they do not represent real resource costs such as those described in the preceding paragraph and are not commensurable with other efficiency costs and benefits.[4]

Another measure of impact is the change in the balance of payments due to reduced exports or increased imports. But again, the relationship between this impact measure and the real changes in availability of goods and services is complex and depends on the particular circumstances.

There are two points to be made concerning these noncommensurate dollar impact measures. First, if they are included in the information

[4]See Freeman (1974) for a further discussion of adjustment costs and assistance policies.

presented to the decision maker, it should be clearly indicated that they cannot be added to economic efficiency benefits or costs without an explicit value judgment concerning the weights to be attached. And second, it is preferable to make the effort to convert impact measures into efficiency benefits and costs by using techniques derived from the logic of applied welfare economics and benefit–cost analysis.

Other Noncommensurables

If market prices and shadow prices are fully utilized to value economic efficiency effects, the initial list of noncommensurate effects of the decision will have been reduced to:

a. fully commensurate economic efficiency benefits and costs measured in dollars; and

b. noncommensurate effects, described and quantified in other units, of which the most significant are likely to be hazards to health and life, damages to the environment and ecosystems, and the distribution of benefits, costs, and hazards among individuals and groups.

This is about as far as the analyst can go in summarizing basic information for the decision maker without imposing value judgments of his own. The ultimate responsiblity for making the necessary trade-off decision, for assigning subjective values, lies with the decision maker. However, the analyst may still be of some assistance to the decision maker.

Consider, for example, the hazard–value-of-life problem. There is a substantial body of information concerning the values placed on human life by other public decision makers as well as those revealed by individuals as they respond to different risky situations and economic signals.[5] These revealed values are inconsistent and contradictory, and there are severe conceptual, theoretical, ethical, and empirical problems in interpreting the available information. A summary and interpretive discussion of this information could be provided the decision maker, not as an explicit part of the decision analysis but as a side display, so that the decision maker can improve his understanding of the issues he confronts. Provided that the decision maker does not feel bound by this information but rather tries to learn from it, he is likely to make better decisions.

[5]See, for example, Schelling (1968), Mishan (1971), Freeman (1973), and Campbell (1974, especially pp. 20–21) for examples and for discussion of the conceptual basis for valuing life. See also Starr (1972) for a discussion of revealed trade-offs between risk and benefit.

In the case of distributional questions, the analyst might be able to assist the decision maker after finding out what dimensions of equity are viewed as most important to the decision maker. The decision maker might be concerned with the distribution of economic magnitudes across regions or across income classes. Or he could be concerned with intergenerational impacts. The analyst could show the decision maker that value judgments about income distribution can be encoded in a system of weights based upon notions of deservingness (Freeman 1969), and the analyst could then show the implications of alternative value judgments.

VALUATION THROUGH TIME AND INTERGENERATIONAL WEIGHTS

In traditional benefit–cost analysis, the social rate of discount is used to weight benefits and costs occurring at different points in time so as to make them commensurable. For example, a dollar of benefit arising 75 years in the future would be equivalent to about 3 cents arising today if the social rate of discount were 5 percent. Because of their persistence, mobility, and environmental accumulation, some of the new chemical substances being introduced carry with them the possibility of irreversible changes in the environment and other long-term effects which could substantially alter the benefits, costs, and perhaps the choices available to future generations.

There have been long-standing debates as to the appropriateness of applying a discount rate to effects on future generations, since any positive rate of discount will directly discriminate in favor of choices that involve bad impacts on later generations but not on earlier ones. Again by way of example, if the discount rate were 5 percent, 100 cases of toxic poisoning 75 years from now would be equivalent to about 3 cases today; or 1 case today would be valued the same as 1,730 cases occurring in 200 years, or the same as the current world population (more than 3 billion cases) in 450 years. Clearly, intergenerational effects of these magnitudes are ethically unacceptable; yet they might be made to appear acceptable if the traditional social rate of discount concept were used to discount future costs to compare with present benefits. Some other method of ethically weighting intergenerational incidence of effects must be devised.

Should all impacts that are intergenerational be weighted equally? That is, should a poisoning case occurring in 75 years, 750 years, or 7.5×10^{12} years be given the same weight as a case today? It might be anticipated that society would have evolved so substantially in hundreds or thousands of years that an equal valuation would be unfair to the present. But this argument really reflects uncertainty about the magnitude of future benefits and costs and how they should be measured rather than about how various future generations should be weighted or compared. In other words, the ethical question of how to weight future generations is not the same as the empirical question as to what their preferences and technologies, and, thereby, their benefits and costs are likely to be.

If there is complete ignorance as to what future preferences will be, it would appear to be defensible to assert that preferences of future generations will be the same as ours. If their preferences are the same, then we have made a correct decision. If their preferences turn out to be different, at least it can be asserted that the present generation did not knowingly discriminate against them. But given this tentative decision for evaluating future preferences, should a positive rate of discount be applied to reflect our uncertainty about their preferences? Several researchers have shown that under rather general circumstances, if the present generation desires to minimize the regret of future generations as to the present generation's choices, then the appropriate social rate of discount is zero, i.e., all generations would be valued equally over a *finite* planning horizon (Schulze 1974).

Whether minimizing regret is an adequate criterion acceptable to the present generation has not been operationally tested. Thus we must tentatively conclude that there is no generally accepted method to weight intergenerational incidence of benefits and costs. Until one is derived, it would appear appropriate to formulate a set of guidelines for valuation that span at least many of the possibilities for deriving intergenerational weights. One such guideline would be to use sensitivity analysis and to compute benefits and costs over a range of social rates of discount, *including a zero rate,* and to see if the outcomes over the range substantially influence the magnitude of benefits in relation to costs.

DECISION ANALYSIS AND UNCERTAINTY

Decisions on controlling toxic substances would be difficult even if the consequences could be predicted with certainty, because those responsible for the decision must address difficult trade-offs between health, environmental effects, and economic values. Most of the time the consequences of the decision are not certain. The effect of regulatory control on dissemination of the substance in the environment, its toxic properties, and the viability of substitute materials are typically not well understood. These uncertainties make the decision process more complex and difficult, and in such situations an explicit framework for the decision may be useful as an aid to those responsible for the decision.

Methodology is available to incorporate uncertainty into a benefit–cost analysis. The concepts, derived from economics, systems analysis, probability theory, and decision theory, are relatively well known; many, in fact, are hundreds of years old.[6] However, their application to complex issues of public policy has been relatively rare. Some examples of analyses that have been carried out on toxic substance regulation and similar public policy issues will be discussed below.

The approach requires technical skill plus an ability to communicate well with the decision maker and the experts on whom he relies for information. The analysis does not provide a method to circumvent difficult judgments. In fact, it may focus attention on issues that have previously been avoided. The process of stating judgments on values and uncertainty in explicit form may be discomforting to many in the decision-making process. Yet a decision maker who is sincerely concerned about achieving insight into situations where a complex decision must be made on the basis of limited data usually finds the analysis to be of considerable assistance.

THE DISTINCTION BETWEEN A GOOD OUTCOME AND A GOOD DECISION

At the outset, the goals and limitations of the analysis must be understood. Formal analysis cannot guarantee that the consequences of the decision maker's choice will be favorable. The analysis does not avoid the need to make decisions whose consequences are uncertain; this is the reality that confronts the decision maker with or without formal analysis. Whereas the decision maker would like to achieve a good outcome, he does not have direct control but must choose among decision alternatives with uncertain consequences. In this situation, the best he can do is make decisions that are consistent with the information available, his values, and his options. The goal of the analysis is to assist the decision maker in making a decision that is logically consistent with what he knows, what he wants to achieve, and the alternatives available to him.

[6]The body of concepts and methodology for analyzing complex decisions in the face of uncertainty is often referred to as decision analysis. For a more detailed discussion, see Howard (1966, 1968), North (1968), Raiffa (1968), and Tribus (1969).

PROBABILITY AND DECISION ANALYSIS

The approach taken in a decision analysis is to describe the consequences of a decision in two dimensions. One is information: if a particular option is chosen, what are the possible consequences? How likely is each of their possible outcomes? Those questions will be answered in terms of probabilities. The other dimension is that of value judgment: what is a particular outcome or set of consequences worth? The process of making value judgments was described in the last section. By asking the decision maker to assess trade-offs, we attempt to describe all of the elements of value in terms of a single commensurate scale, usually monetary value. Assessment of these trade-offs is often the most difficult and important part of the analytical process.

A simple example will illustrate. Suppose a decision must be made between banning and permitting the use of a pesticide. If the pesticide is banned, there will be no human health effects; but if use is permitted, an uncertain number of deaths from cancer may result. This uncertainty can be quantified by assigning probabilities to each of the possible outcomes for the consequence of concern, i.e., deaths from cancer. For example, it could be stated that there is a probability of 90 percent that no additional deaths would result from use of the pesticide; a probability of 5 percent that about 10 additional deaths would result in the population at risk; and a probability of 5 percent that about 100 additional deaths would result. A more detailed description of the uncertainty would be given by a probability distribution over a continuous range of outcomes, i.e., all the levels of cancer mortality that might occur. We will use the three discrete outcomes above to illustrate how probabilities are used in calculations.

For many decisions it is appropriate to use the average or expected value of the outcome as a decision criterion. The expected value is computed by multiplying the probability of each outcome times the magnitude of that outcome and summing this product over all the possible outcomes. In the example above, the expected mortality from using the pesticide is

$$\begin{aligned} E[V] &= 0.90 \times 0 + 0.05 \times 10 + 0.05 \times 100 \\ &= 5.5 \text{ deaths.} \end{aligned}$$

If the decision maker assigns a monetary value to each cancer death of \$500,000, then the expected value of the loss associated with pesticide use is 5.5 × \$500,000 = \$2.75 million. This expected loss would then be compared to other important consequences, e.g., the economic loss that would accrue if the chemical were banned.

RISK PREFERENCE

When values in question are of moderate size, the criterion of maximizing expected monetary value is generally adequate. If there is disagreement on the decision criterion, it can usually be traced to the assessment of trade-offs and the assigning of values under certainty, e.g., the value to be assigned to a death from cancer.

However, for some situations, expected value is clearly inadequate as a criterion. Suppose that continued use of an aerosol propellant is judged to have a probability of 0.001 of causing significant depletion of the ozone layer, with massive consequences for skin cancer, diminished crop yields, and environmental damage. These consequences are valued as a social loss of approximately $1 trillion. Thus, the expected value of the loss from continued use of the chemical is $1 billion. The expected economic loss if the chemical were banned is estimated to be $10 billion. Based on a criterion of maximizing expected value, the continued use alternative appears preferable. However, many decision makers might feel that because of the enormousness of the possible consequences, the decision should be to ban aerosols. Even though the expected value is small, we "cannot afford the risk."

Von Neumann-Morgenstern utility theory provides a general approach to evaluating uncertain situations, such as those where expected value is deemed an inadequate means of describing the range of possible outcomes (Von Neumann and Morgenstern 1947).[6] The usefulness of this theory is that it provides a means of assessing the decision maker's attitude toward risk in quantitative form, and then using this assessment to evaluate choices between complex uncertain situations. In practical applications, however, expected value is probably adequate as a criterion for government policy decisions unless there is a possibility of disaster affecting a large portion of the nation's population or resources. The quantification of probabilities and values is generally a much more difficult aspect of the analysis than assessing a decision maker's risk attitude. The sensitivity of the decision to risk attitude can be ascertained to see if this aspect of the problem deserves further attention (Howard 1968).

[6]The body of concepts and methodology for analyzing complex decisions in the face of uncertainty is often referred to as decision analysis. For a more detailed discussion, see Howard (1966, 1968), North (1968), Raiffa (1968), and Tribus (1969).

PROBABILITY AS A LANGUAGE FOR JUDGMENT ABOUT UNCERTAINTY

Most people are accustomed to using probability informally as a language for describing uncertainty. Tomorrow's weather is described in terms of probability of rain, and a statement that some event is uncertain often elicits the question, "What are the chances it will come out well—or badly?" Probability theory provides an unambiguous and logically consistent way to reason about uncertainty. In fact, a forceful argument can be made that any logical process for reasoning about uncertainty is equivalent to probability theory (Cox 1961, Raiffa 1968, Savage 1954, Tribus 1969).

The same people who use probability naturally in informal situations may be extremely reluctant to use the same ideas in important professional decision situations. To some extent the problem is one of measurement. Many people are accustomed to viewing probabilities as exact numbers based on objective evidence, and they are uncomfortable with the idea of using probabilities as a language for expressing their judgment on a matter where the available information is limited.

Textbooks on probability usually take examples in which the probability values are deduced from physical symmetry. If we flip a coin, what is the probability that the head will turn up three times in a row? "One-eighth," we reply—based on the assumption that since the coin is symmetrical both sides are equally likely to come up. Thus, the probability of heads is one-half on each flip, and three heads in three tosses has a probability of $(1/2)^3$, or $1/8$. To test the importance of the symmetry assumption, consider flipping a thumbtack: what probability would you assign to its landing with the point up three times in a row?

Another way to assign probabilities is on the basis of past frequencies. The thumbtack problem is not difficult if we have flipped a thumbtack 10,000 times and observed the results: we just use the frequency with which the point-up outcome has been observed for the probability that the tack will land point-up on the next flip. But the usual situation is that we do not have the data. We have not flipped thumbtacks and recorded the experimental results. Similarly, if we had statistical data on incidence of disease given dosage, it would be easy to assign a probability to an individual's becoming ill as a result of ingesting a given amount of a chemical. But the typical situation is that decisions are required before this type of data is available. Judgment must be extrapolated from other information, for example, animal tests or dose-reponse relationships for chemicals of similar structure. In some situations the expert may be confident in his judgment in the sense that he would not expect it to change if he could obtain further information. In other situations further information might be expected to change the judgment a great deal. For example, we might feel that an opportunity to observe 10 flips would greatly improve our ability to make judgments about the outcome of three tosses of a thumbtack.

The probability theory for problems such as the thumbtack and environmental uncertainties is relatively straightforward, once the basic assumption has been accepted that probability assignments are not "objective" but represent judgments, that they summarize information or a state of mind rather than being physically measurable.[7] Where symmetry considerations are used to assign probabilities, there is always a judgment that various possibilities (e.g., heads or tails) are equally likely. Where past frequencies are used as the basis for probability assignments, the judgment has been made that the present situation to which a probability is being assigned is equivalent to past situations in which the frequency data was observed. This judgment may not be warranted. For example, toxic substances often have synergistic effects. A poor prediction may be the result of extrapolating frequency data on the toxicity of chemical A from a situation where chemical B was not present to a situation in which substance B is present. There are no simple formulas from statistics or probability theory that allow a scientist to avoid this sort of mistake. He must use his judgment to relate past observations and other relevant information to uncertain events in the future. Probability theory can provide him with some assistance in applying his judgment consistently, but it cannot provide a substitute for good judgment.

Once the basic premise has been accepted that probabilities are not limited to situations in which frequencies (e.g., "good data") are available as a basis for assigning them, we can then use probability theory to reason about the full spectrum of uncertainties that may result from decision alternatives. The probability, for example, that continued use of an aerosol propellant will affect the ozone layer must be taken as scientific judgment: nothing more or nothing less. The probability value acts as a summary of the scientific judgment of experts in a form useful to the policy maker who must decide whether to ban this chemical.

The difficulty of encoding such judgment in the form of a probability is not trivial. Most people do not find it easy to express their judgment consistently, and both skill and judgment are required on the part of the

[7]For further discussion of probability and thumbtacks, see North (1968), Judd et al. (1974), or Howard (1970). This problem is an example of the Bayesian viewpoint on probability, as opposed to the more limited viewpoint of traditional statistics that probabilities must be frequencies.

analyst for the encoding process.[8] If the probabilities are small, the difficulties are even more severe: even trained scientists do not generally have intuition about the difference between one chance in a hundred and one chance in a thousand. Often it is useful to model the relationships that lead to the event of interest. Complex models have been used in this fashion to develop probability assignments for improbable environmental disasters, such as accidental release of radioactivity from a nuclear reactor (U.S. AEC 1974) or contamination of another planet by terrestrial microorganisms transported aboard a spacecraft (Judd et al. 1974).

THE VALUE OF RESOLVING UNCERTAINTY

How can a decision maker evaluate whether more information should be obtained? An important aspect of a decision analysis approach is that it facilitates the evaluation of alternatives to resolve uncertainty. Often a crucial question on toxic substances is whether a substance should be restricted or banned on the basis of limited available information, or whether continued use should be permitted while more information is collected. The method for evaluating the worth of additional information is simple in concept. We evaluate the expected impact of the information in changing the decision.

Suppose a decision must be made on a potentially toxic chemical in wide use. With probability 0.10 this chemical is estimated to cause health damage valued at $15 billion. The economic loss from a decision to ban the chemical is estimated to be $1 billion. If a decision had to be made on present information, the best alternative would appear to be banning the chemical, since the expected health damages (0.10 × 15 = $1.5 billion) outweigh the economic losses. But suppose the uncertainty could

[8]The usual method is to ask the subject whose judgment is being assessed to state the probability (or odds) at which he would be willing to bet on the occurrence of the uncertain event in question. One of the difficulties in assessing judgment in this manner is that people often have a distinct preference for betting on situations where probabilities can be deduced from frequencies or symmetry con- siderations. For example, many people will prefer to wager on the call of a coin rather than the call of a thumbtack, although the asymmetrical tack should give them at least as high a chance for making the winning call as the symmetrical coin. There is much literature on this problem, which is sometimes referred to as the "Ellsberg Paradox" (Ellsberg 1961, Fellner 1961, Raiffa 1961, and a number of subsequent articles in the *Quarterly Journal of Economics*). The difficulties of assessing human judgment on uncertain events have been extensively studied, and much is known about the types of errors that can occur in the encoding process (Tversky and Kahnemann 1974). A recent review on practical methods for encoding probability distributions is found in Spetzler and Staël von Holstein (1975).

be resolved before the decision was made. With probability 0.90 the information would show the chemical was nontoxic, and the billion dollar (economic) loss could be avoided. The expected improvement from the information is 0.9 × $1 billion or $900 million. If uncertainty can be resolved in a one-year research program and if the expected health damages from one year of continued use are judged to be well below $900 million, it may be advisable to defer the decision to ban the chemical for a year. Conversely, if the expected health damage from another year's use is well above $900 million, the best course of action may be to ban the chemical on the basis of the limited information that is summarized by the probability of 0.10 that the chemical has major toxic effects.

Calculations on the value of resolving uncertainty are easily carried out once a decision problem has been formulated in terms of explicit values and probabilities. These calculations on the value of additional information may be very useful in setting research priorities and allocating research budgets.

LEGAL CONSTRAINTS

We have spoken of probability in the mathematical sense as a language for describing uncertain situations. However, there appear to be legal impediments to application of this language in some areas of EPA's regulatory authority.

In two recent cases, judges have apparently interpreted the term probability in a different sense, and this may create obstacles to the use of probability concepts in regulatory decisions which are subject to judicial review. In the first case (*United States* v. *Reserve Mining Company,* Eighth U.S. Circuit Court of Appeals, June 4, 1974), the court ruled that the government could not impose regulations on an industry based upon a less than certain health risk when a certain adverse economic impact would occur from the decision. In the second case (*Ethyl Corporation* v. *Environmental Protection Agency* U.S. Court of Appeals of the District of Columbia, January 1975), the court ruled that the Administration acted improperly in attempting to regulate a chemical (the lead content of gasoline) based upon a less than probable (50 percent) association with a major health effect (lead poisoning).

In some instances a comprehensive risk–benefit assessment as a basis for policy is at variance with the law. For example, the Clean Air Act of 1970 dictates that only health and related risks be considered in establishing primary standards for air quality.

IMPLEMENTATION

The aspects of decision analysis described above can be combined and implemented in a systematic approach to the problems of making regulatory decisions about chemicals in the environment. Examples of the application of this approach to the regulation of toxic substances in the environment include a recent National Academy of Sciences (NAS) report that addressed the control of sulfur oxide emissions from coal-burning power plants (North and Merkhofer 1975) and a Stanford Research Institute project for EPA currently in progress on determining acceptable risks from cadmium, asbestos, and other hazardous wastes (Moll et al. 1975). Applications to similar public policy problems include deployment of weather modification technology (Boyd et al. 1971, Howard et al. 1972) and regulation of hazardous consumer products such as flammable fabrics (Tribus 1970).

The procedures and emphasis in applying benefit–cost analysis may vary a great deal, depending on the character of the problem and the level of analytical resources available for the application. However, the following steps may serve as a general outline and guide for the approach.

1. Specify the decision or decisions to be addressed by the analysis, and develop the analysis to give insight into those decisions.

The aim of the analysis is to assist a decision maker in choosing among alternatives for controlling hazardous substances. Before embarking on extensive information gathering or modeling efforts, it is important to specify the substances and the alternative regulatory and control mechanisms to be examined. The evolution of the analysis can then be guided by the sensitivity of the decision maker's best choice to the various issues involved. If, for example, different assumptions about the dose–response relationship for the health effects caused by ingesting a toxic substance would lead to different preferred control strategies, considerable effort will be appropriate in assessing the dose–response relationship. However, for some toxic substances, damage to materials and nonhuman species may motivate strict control for reasons exclusive of human health, and the dose–response relation will have little or no impact on the decision among regulatory alternatives. Then little analytical effort will be appropriate for assessing the human health dose–response relationship. The analysis should maintain a focus on the decision problem confronting the decision maker, and it should strive to identify which among the many complex issues are most important in determining the best decision alternative.

2. Review the available background information on the past and present justification for standards applicable to the chemical of interest.

Some perspective on the emergence of concern and the rationale for selection of the chemical as potentially hazardous should be obtained in this initial investigation. This step, and the subsequent ones, should vary in depth according to the problem and to the time and effort available. Detailed and comprehensive investigation may be appropriate in a problem involving an initial standard to be set on a major environmental problem such as sulfur oxide emissions. Conversely, a brief review of available information may be all that is appropriate for a problem involving only small changes in existing standards or applications, such as one of revising cost data to account for inflation. But, gathered to whatever depth, the information collected in this initial investigation should establish a base level upon which the following steps can be built.

3. Describe the flow of the chemical from production through refining and use to eventual disposal.

This description should be quantified to show the materials balance from stage to stage and in particular the escape of emissions into the environment at each stage. The study of cadmium standards now being done for EPA as a demonstration of standard-setting methods illustrates how such an analysis can account for the major emissions of a hazardous material (Moll et al. 1975). In addition, monitoring information on emissions at the various stages is needed both as input and as a follow-up verification of the estimates.

4. Trace the flows of the chemical from points of emissions into the various media of the environment that lead to exposure to people and adverse effects on human health, damage to susceptible species, or other detrimental effects.

This step also requires two types of information inputs. A predictive model must enable the analyst to estimate how and over what time period the emission will disperse into the environment. This model should be linked as much as possible to actual observation by means of field monitoring data. The monitoring data should be sensitive enough to distinguish between contaminant and background levels, and to verify chemical transfers from one environmental medium to another (e.g., air to ground). Where field data are inadequate, the analyst must generate a plausible model of the dispersal process (supported by available data from experiments, observed analogous processes, and so forth) in order to establish this link in the chain of logic.

It is at this step that another ingredient of the method should be introduced, viz., modeling of the uncertainty. Obviously, no model will be able to account accurately for the emissions and dispersal of a chemical, so estimates must be stated in probability terms or with some form of sensitivity limits attached. These limits will vary depending upon the chemical, the dispersion conditions, and the spatial and temporal pattern of emissions. The sulfate conversion model in the NAS study (North and Merkhofer 1975) illustrates how limited data may be used to develop a probabilistic description of the relationship between emissions and ambient concentrations.

5. Estimate the effects of exposures on people and the environment. Effects on human health can be described by means of a dose–damage function with uncertainties properly represented (Moll et al. 1975, North and Merkhofer 1975). Merely specifying a threshold level or a median lethal dose is not adequate, because these measures do not indicate sensitivity of the damage to the dose level. A dose–damage function can usually be derived even when data are uncertain or conflicting, because explicit recognition can be given of the range of uncertainty.

When the dose–damage relation is uncertain, probabilities can be assigned over the range of possible increases in health effects for a given increase in dose level. This approach has much to recommend it over attempting to assess a safety margin, as has been pointed out in a previous NAS report (Palmes 1974). When epidemiological data is scant or nonexistent, it is rarely possible to state categorically that dosage below any specific level is "safe." It seems preferable for those responsible for regulatory standards to state as clearly as possible that there are possible detrimental consequences, what these adverse consequences might be (e.g., eye irritation, cancer, and others), and what the probability is that the consequences will occur with a given dosage. These assessments should be supported by citing the information on which they are based, and ranges for sensitivity analysis should be provided. Such assessments can serve not only as a summary of information for regulatory decisions, but also as a basis for setting research priorities for animal and epidemiological studies to resolve the uncertainties in the dose–damage relationship.

Effects other than on human health (e.g., environmental effects) are often considerably more difficult to estimate because the relevant information has not been collected. Depending on the type of environmental concern, one might measure effects in terms of contaminant levels in air or water, ecological "energy balance," numbers of lost animals or acres of viable ground, reduced recreational or real estate potential, and so forth.

6. Specify the costs and effectiveness of alternative controls that might be applied to the use of the chemical to reduce human or environmental exposure levels.

Costs will include the combination of capital, operating, and enforcement costs described in an earlier section; and effectiveness will be in terms of an absolute or percentage decrease in the exposure levels achieved by the control alternative. Both costs and effectiveness may be uncertain, and it may be appropriate to describe these uncertainties by using probability distributions. On occasion, a full analysis of control alternatives may not be required when the logic or the evidence indicates that only one alternative to controlled use is feasible, or that none is.

7. Estimate the economic costs of control and/or reductions in economic benefits associated with imposing alternative regulations or controls on the production and use of the chemical.

Control costs may be derived from engineering estimates. The costs associated with restrictions on use may be based on estimates of the costs of substitute products or on estimates of lost consumer and producer surpluses derived from demand curve analysis (Freeman 1973, Moll et al. 1975).

Economic effects should be expressed as a function of time. Limits for sensitivity analysis can be established for the elasticities of supply and demand and for projected future magnitudes in order to assess the sensitivity of the impacts to changes in the economic assumptions. To aid in subsequent distributional analyses, the economic as well as hazard impacts can be disaggregated in terms of areas, groups, and time periods involved. Related but not commensurable effects include the costs of plant shutdowns and effects on the balance of payments.

8. Summarize and array the results of the analysis for presentation to the decision maker.

A suggested format for summary display is shown in Figure H-1. Without attempting to be inclusive, we show in this display matrix how such a summary might be arranged to consider questions of banning a chemical. It permits the analyst to summarize the major factors affecting the decision, and does it in a manner that retains a degree of flexibility. The comments column permits the analyst to call attention to those considerations that may be crucial in any particular case, such as sensitivity of the dependent variables to input uncertainties, relationship of hazards and benefits to the populations involved, and the time frame used for the analysis.

It may be useful to extract the most important elements of the analysis for visual display to the decision maker. Figure H-2 illustrates such a display. It shows the impacts of two alternative control plans in terms of both health or hazard effects and economic costs and benefits.

Net economic costs include all real resource costs of control and economic efficiency benefits forgone due to control, all defined and

EFFECTS	COMMENTS (Quality and extent of data, uncertainties, other concerns)	ALTERNATIVES Ban Now	 Ban Phase Out	 Control A	 Alternatives B
I. Hazards Avoided A. Health 1. Lost person years 2. Lost activity days 3. Population exposed					
B. Environmental 1. Material damaged 2. Vegetation damaged 3. Animal losses 4. Aesthetics 5. Recreation					
II. Benefits Lost 1. Net Market Value 2. Years extended life 3. Consumer benefits 4. Aesthetic improvement					
III. Cost of Control 1. Capital 2. Annual operating 3. Implementing costs					
IV. Economic Impact 1. Plants closed 2. Jobs lost					
V. Distributional Effects 1. Local-now 2. National-now 3. National-50 years 4. Population Group A 5. Population Group B					

FIGURE H-1 Display of major effects of control on hazards, costs, and benefits.

measured in commensurate dollar terms. Both health hazards and economic costs are represented as incremental changes from the present situation. Uncertainties can be portrayed as confidence limits derived from probabilistic statements discussed above. This two-dimensional display assumes, for simplicity, that the other effects of control (e.g., environmental, distributional) which are summarized in the matrix display are either of second-order magnitude or unimportant to the decision maker.

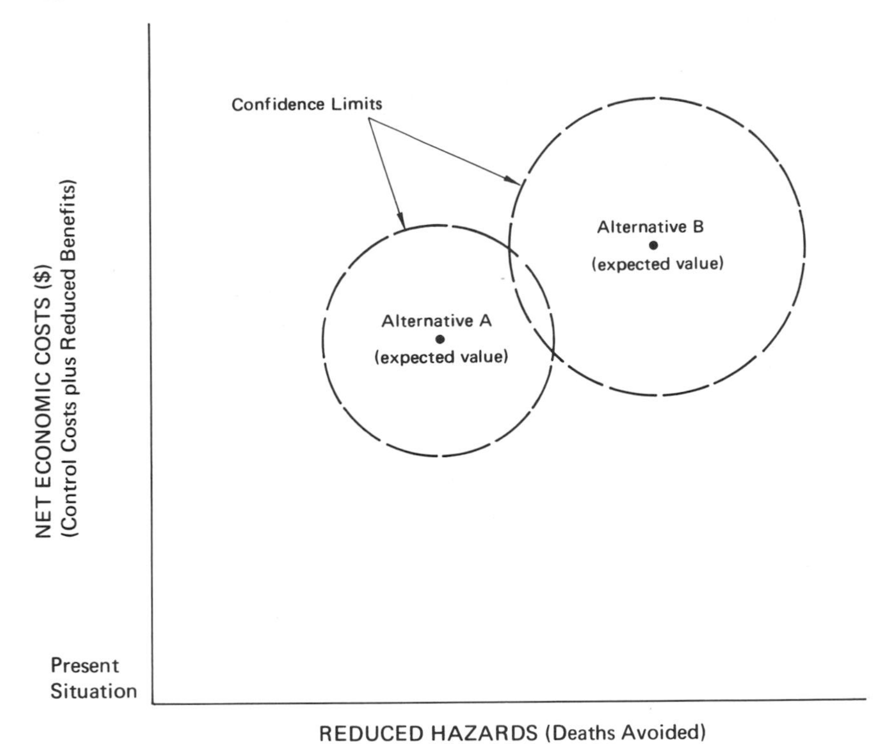

FIGURE H-2 Example of a hazard–cost–benefit display.

Ultimately, the choice among control alternatives—or the choice of whether to control at all—depends on the relative values assigned to economic magnitudes and reduced hazards. In two-dimensional displays such as Figures H-2 and H-3, alternative value judgments about the trade-off between these two magnitudes can be represented by a family of straight lines called *indifference lines* whose slope is equal to the value one wishes to attach to human life. The name for these lines stems from the fact that given the value of life, a decision maker would be indifferent to any two alternatives which were on the same line (see Figure H-3). For this figure, the solid lines labeled "$/Hazard" represent one possible value relationship between dollars and deaths avoided. The more steeply sloped, dashed lines portray a *higher* dollar value per death avoided.

The objective of the decision maker is to choose that alternative which lies on the indifference line closest to the lower right-hand corner of Figure H-3. This is because lines below and to the right represent both

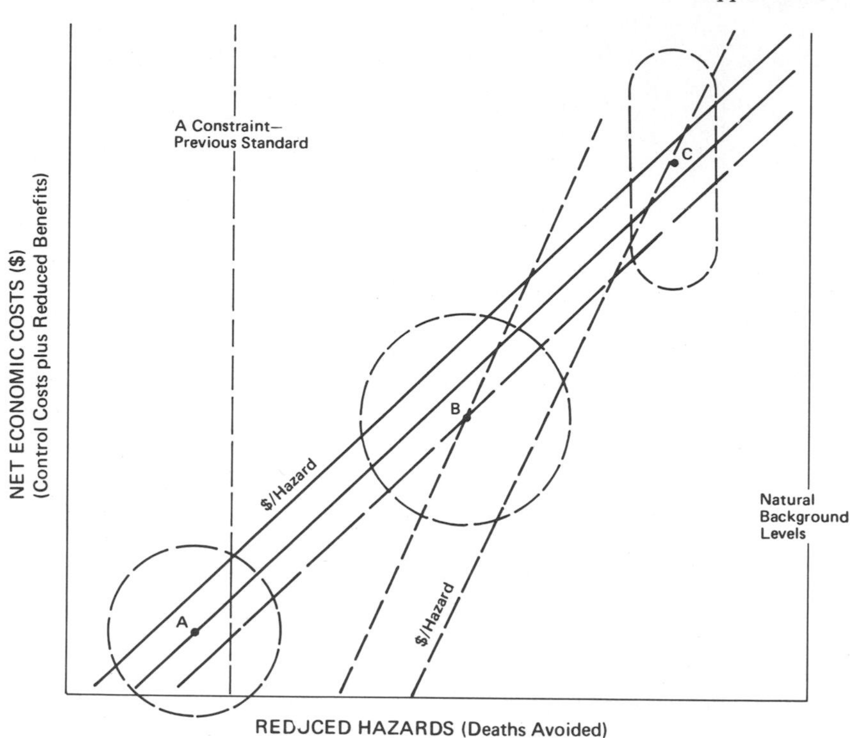

FIGURE H-3 Value judgments in the hazard–cost–benefit display.

fewer deaths and smaller reductions in benefits. Given the value judgment concerning the value of deaths avoided which is portrayed by the solid indifference lines, Alternative B is preferred. However, the implications of a higher value for deaths avoided can easily be portrayed. A higher dollar value means a more steeply sloped family of indifference lines. At least for some higher dollar values, such as that portrayed by the dashed indifference lines, Alternative C would be preferred. The visual displays of Figures H-2 and H-3 make it easier for the decision maker to see the implications of alternative value judgments, and they may facilitate his consideration of the interaction between choice and value.

The background level of pollution and its associated health hazard represents a lower level of hazard reduction which cannot be surpassed. This constraint (maximum attainable reduction in hazard) is shown as a vertical line in Figure H-3. Sometimes an upper level of permissible hazard may also be introduced because society is not willing (either currently or in some long-term sense) to accept a group hazard greater than some level. Preexisting standards or accepted guidelines represent such a constraint.

To reflect distribution considerations, i.e., who gets what benefits and who is exposed to what hazards, alternatives shown on the hazard–benefit chart can be calculated for different interest groups. For example, Alternative A, as shown, might represent the trade-off for a particular local area, and a different point might represent Alternative A for the U.S. as a whole. When interest-group considerations are shown in this way, the decision maker can make his own judgment of the appropriate weightings for the two trade-offs. Other distributional trade-offs, e.g., the intergenerational problem, can be handled in an exactly analogous manner.

Many of these considerations will have ranges arising from uncertainty or differences in value judgments. Illustration of these ranges on the hazard–benefit chart may help to define the domain of feasible solutions in standard setting. The EPA cadmium study shows how these types of considerations can be used in a hazard–benefit chart of an actual standards problem (Moll et al. 1975).

The decision maker can then incorporate into his own deliberations what he considers to be the most pertinent considerations from these displays and the accompanying documentation. He will then be able to interpret the implications of his own value judgments regarding such things as the value of human life, the value of avoiding risk, or distributional objectives for the decision he is about to make.

9. Use the same kind of step-by-step analysis and graphical display to indicate the types of information needed to narrow the uncertainties for future decisions.

This step provides a systematic method of planning for future data integration at the same time as efforts are underway to make decisions on the basis of inadequate present data. The analysis of sulfur oxide emissions illustrates a way to assess the expected value of further information (North and Merkhofer 1975). Another illustration of this type of assessment is found in Howard et al. (1972).

CONCLUSIONS AND RECOMMENDATIONS FOR CONSIDERATION BY THE COMMITTEE

In this report we have not attempted to present a comprehensive and rigorous development of the theory of benefit–cost analysis and decision theory, although some references to the relevant literature have been provided. Neither have we presented a cookbook full of techniques and practitioner's tricks of the trade. Nor have we dealt with any of a number of unresolved issues and problems in the theory and application of decision analysis (but see Lind 1972).

We have described benefit–cost and decision analysis as a useful framework for dealing with the kinds of problems created by the necessity for making regulatory decisions about chemicals. We regard this framework, in part, as a way of organizing information; in part, as a mode of thought which is essentially analytical and rational/logical; and, in part, as a language which permits clear and consistent expression of difficult ideas and facilitates discussion between the decision maker and analyst.

The framework discussed here can be incorporated into an ongoing decision-making process. Where a sequence of decisions must be made over time (e.g., deciding whether a chemical should be investigated, setting interim standards, final standards, revision, and so forth), the framework can be used to organize the available and relevant information at each decision point. By providing a consistent way of organizing and presenting information, it facilitates the incorporation of new data as they become available.

Finally, as a mode of thought, the framework can be used at many levels, from back-of-the-envelope "quicky" decisions (e.g., whether or not to investigate a problem more thoroughly) to large-scale formal analyses, including, perhaps, computer-based models and simulation systems. Yet at whatever level or scale it is used, it remains simply a framework and not a formula, making more possible the reasonable execution of politically responsible judgment.

We close with two recommendations.

First, we recommend that EPA adopt the principles of benefit–cost and decision analysis as a basis for organizing and displaying information, and that it direct staff personnel to prepare more detailed guidelines concerning procedures and techniques.

Second, there are, currently, legal constraints which prohibit the use of risk–benefit assessments for decision making under certain laws. Similarly, in the instance of a low probability of a high-cost risk, recent court decisions have made it difficult to use the language of probability in risk–benefit comparisons. Therefore, we recommend that, if a benefit–cost or decision theory framework is to be used, the legal framework under which these decisions are made be altered.

REFERENCES

Boyd, D. W., R. A. Howard, J. E. Matheson, and D. W. North (1971) Decision Analysis of Hurricane Modification. Final Report, SRI Project 8503, prepared for Techniques Development Laboratory, Systems Development Office, National Weather Service, National Oceanic and Atmospheric Administration, U.S. Department of Commerce, Washington, D.C. (Available from National Technical Information Service as #COM-71-00784).

Campbell, R. R. (1974) Food Safety Regulation: A Study of the Use and Limitations of Cost–Benefit Analysis. Washington, D.C.: American Enterprise Institute.

Cox, R. T. (1961) The Algebra of Probable Inference. Baltimore: Johns Hopkins University Press.

Ellsburg, D. (1961) Risk, Ambiguity, and the Savage Axioms. Quarterly Journal of Economics 75(4): 743–669.

Fellner, W. (1961) Distortion of Subjective Probabilities as a Reaction to Uncertainty. Quarterly Journal of Economics 75(4): 670–689.

Freeman, A. M., III (1969) Project Design and Evaluation with Multiple Objective. In U.S. Congress, Joint Economic Committee, Subcommittee on Economy in Government, The Analysis and Evaluation of Public Expenditure: The PPB System. Washington, D.C.

Freeman, A. M., III (1973) A Survey of the Techniques for Measuring the Benefits of Water Quality Improvement. Prepared for the USEPA Symposium on Cost–Benefit Analysis in Water Pollution Control.

Freeman, A. M., III (1974) Evaluation of Adjustment Assistance Programs with Application for Pollution Control. U.S. Environmental Protection Agency, Washington, D.C.

Haveman, R. H. and B. A. Weisbrod (1973) The Concept of Benefits. Prepared for the USEPA Symposium on Cost–Benefit Analysis in Water Pollution Control.

Howard, R. A. (1966) Decision Analysis: Applied Decision Theory. In Proceedings of the Fourth International Conference on Operational Research. New York: Wiley-Interscience, pp. 55–71.

Howard, R. A. (1968) The Foundations of Decision Analysis. IEEE Transactions on Systems Science and Cybernetics. SSC-4(3).

Howard, R. A. (1970) Decision Analysis: Perspectives on Inference, Decision and Experimentation. Proceedings of the IEEE 58(5).

Howard, R. A., J. E. Matheson, and D. W. North (1972) The Decision to Seed Hurricanes. Science 176: 1191–1202. June 16.

Judd, B. R., D. W. North, and J. Pezier (1974) Assessing the Probability of Contaminating Mars, Final Report. SRI Report MSU-2788, Stanford Research Institute, Menlo Park.

Lind, R. C. (1972) The Analysis of Benefit–Risk Decisions. In Perspectives on Benefit–Risk Decision Making, Committee on Public Engineering Policy, National Academy of Engineering, Washington, D.C.

Mishan, E. J. (1971) Evaluation of Life and Limb: A Theoretical Approach. Journal of Political Economy 79(4): 687–705.

Moll, K., S. Baum, E. Capener, F. Dresch, G. Jones, C. Starry, D. Starret, and R. Wright (1975) Methods for Determining Acceptable Risks from Cadmium, Asbestos, and Other Hazardous Wastes. Stanford Research Institute, Menlo Park.

National Academy of Engineering (1972) Committee on Public Engineering Policy. In Perspectives on Benefit–Risk Decision Making. Washington, D.C.: National Academy of Engineering.

North, D. W. (1968) A Tutorial Introduction to Decision Theory. IEEE Transactions on Systems Science and Cybernetics. SSC-4(3).

North, D. W. and M. Merkhofer (1975) Analysis of Alternative Emissions Control Strategies. In Air Quality and Stationary Source Emission Control, a Report by the Commission on Natural Resources, National Academy of Sciences, prepared for the Committee on Public Works, United States Senate. Washington, D.C.: U.S. Government Printing Office.

Palmes, E. D. (1974) Dose–Response Relations. In Air Quality and Automobile Emission Control. A Report by the Coordinating Committee on Air Quality Studies, National Academy of Sciences, prepared for the Committee on Public Works, United States Senate. Washington, D.C.: U.S. Government Printing Office. 2: 479–481.

Raiffa, H. (1961) Risk, Ambiguity, and the Savage Axioms: Comment. Quarterly Journal of Economics 75: 690–694.

Raiffa, H. (1968) Decision Analysis: Introductory Lectures on Choices Under Uncertainty. Reading, Massachusetts: Addison-Wesley.

Savage, L. J. (1954) The Foundations of Statistics. New York: John Wiley & Sons, Inc.

Schelling, T. C. (1968) The Life You Save May Be Your Own. In Problems in Public Expenditure Analysis. Edited by Samuel B. Chase. Washington, D.C.: Brookings Institution.

Schulze, W. (1974) Social Welfare Functions for the Future. American Economist 18(1): 70–81.

Spetzler, C. S. and C.-A. S. Staël von Holstein (1975) Probability Encoding in Decision Analysis. Presented at the ORSA-TIMES-AIEE 1972 Joint National Meeting, Atlantic City, New Jersey, November 1972. To appear in Management Science, 1975.

Starr, C. (1972) Benefit–Cost Studies in Socio-Technical Systems. In Perspectives on Benefit–Risk Decision Making, Committee on Public Engineering Policy, National Academy of Engineering, Washington, D.C.

Tribus, M. (1969) Rational Descriptions, Decision, and Designs. New York: Pergamon Press.

Tribus, M. (1970) Decision Analysis Approach to Satisfying the Requirements of the Flammable Fabrics Act. U.S. Department of Commerce News. February 13.

Tversky, A. and D. Kahnemann (1974) Judgment Under Uncertainty: Heuristics and Biases. Science 185: 1124–1131.

U.S. Atomic Energy Commission (1974) Reactor Safety Study: An Assessment of Accident Risks in U.S. Commercial Nuclear Power Plants. N. Rasmussen, study director. U.S. AEC WASH- 1400.

U.S. Congress (1969) Joint Economic Committee, Subcommittee on Economy in Government. The Analysis and Evaluation of Public Expenditures: The PPD System. Washington, D.C.

Von Neumann, J. and O. Morgenstern (1947) Theory of Games and Economic Behavior, 2nd ed. Princeton: Princeton University Press.

APPENDIX I

Working Paper on Regulatory Options

SUMMARY AND CONCLUSIONS

This paper develops a problem-oriented approach to the decision-making process related to the regulation of chemicals in the environment. Problems requiring decisions may be "crisis" or "noncrisis" in nature, and the implications of each are discussed. A system for information gathering for the decision-making process is constructed. This system incorporates several elements which we believe do not currently exist in the U.S. Environmental Protection Agency (EPA) for the systematic preparation and transmittal through the system of (a) hazards, costs and benefits data, (b) feedback information, and (c) feed-forward information.

Two levels of decision making are discussed: (1) a technically based analytic level where decisions are made within broad guidelines set by the politically accountable decision maker; and (2) a "major" sociopolitical decision level where top-level decision makers participate directly. The panel urges the publication of a "white paper" by the Agency for each major decision describing the rationale that led to the choice of a given regulatory option. The importance of maintaining a clear separation between the roles of technical assessment and sociopolitical decision making is emphasized. We urge specific awareness of the possible influence of past decisions/methodology on present and future decision making.

Next we address evaluation of the possible regulatory options, with all the relevant data displayed for presentation to the decision maker. The dual role of interagency cooperation and competition is explored, and the role of nongovernmental input into this process is reviewed. We recommend that the EPA's Science Advisory Board (SAB) exercise broad, semiautonomous responsibilities in the area of independent peer review of the data base and technical analysis used in the decision-making process. We believe that strong guidelines are necessary to minimize the possibility that such a board approach, free from political accountability, would develop advocacy positions or attempt to make decisions directly.

The authority under which EPA can act is reviewed briefly, particularly those cases where multiple authorities exist or where no authority exists. Some differences between evaluation of premarket and in-use chemicals are presented, and the utility of incremental introduction as a means of preserving incentive for innovation, while minimizing hazard and capital risk, is supported.

INTRODUCTION

The objective of this paper is to describe an analytical process for the resolution of regulatory issues associated with the control of chemicals in the environment. The panel has selected as its method of analysis a "problem-oriented" system, rather than one based solely on reviewing available control mechanisms designed to achieve some preconceived "solution." Thus, we begin with the definition of a generic "problem," determine the information and data-gathering process needed to address the problem, and then examine how the available information should be applied to the problem. Finally, we look at the various alternatives available to confront (or control) the problem and the decisions which must be made to implement the chosen regulatory action. We view this report as an analytical tool, with some specific suggestions which we believe decision makers may find helpful in improving a regulatory agency's organization, information and data-gathering techniques, and decision-making processes related to the regulation of chemicals in the environment.

Two terms require definition at the outset. By "regulatory actions" we mean any action taken under law by an agency which is intended to channel behavior in a socially desirable way. We recognize that this definition is broad enough to include appropriate economic inducements and the marketplace, but we choose to restrict the definition for this discussion to include only those actions which place requirements, limitations, or prohibitions on the conduct (activities) of those subject to federal regulatory authority.

By "chemical pollutant," which may be used interchangeably with "toxic substance," we mean any substance present at levels in the environment which causes concern for health or for the environment.

It may be helpful to state clearly our perception of two relatively distinct levels or categories of decision making within EPA or any other agency concerned with the regulation of toxic substances (or any complex, scientifically and technically based regulatory program, for that matter). Top-level decision makers within EPA and other regulatory agencies generally determine or influence the outcome of regulatory actions in three ways: (1) by setting broad policy direction for agency programs, (2) by reviewing and approving standards and regulations which establish the criteria for specific lower-level decisions, and (3) by the adjudication of cases which are appealed to the head of the agency for final disposition, prior to judicial intervention. In each of these instances, because of the importance of the issues raised, the scope of their impact, or the inability to resolve conflicts at a lower level, the agency head generally is called upon to make sociopolitical value judgments, evaluate trade-offs, or apply some degree of equity and balance to achieve an optimal social benefit. The decision process is familiar to all high-level decision makers, although it is difficult to describe. Most decisions which involve a change in the status quo require this level of decision making, since they either explicitly or implicitly express a new or changed agency policy. A decision to cancel or suspend the registration of a pesticide, to deny a permit to dump hazardous materials into the ocean, or to seek an injunction against discharging hazardous materials into a lake, are examples of regulatory actions requiring the policy evaluation of top decision makers.

The second level or category of decision making is that which involves the application to a particular situation of "decision rules" established by the top-level decision makers. The decision rules may be either published regulations or administrative guidelines subject to the agency's regulatory authority. This encompasses the vast majority of regulatory decisions carried out in a rather predictable fashion pursuant to established criteria, such as the issuance of discharge permits to industries and municipalities, the registration of pesticides, or the specification of emissions limitations for individual stationary sources of air pollution. Also included within this category are a significant but relatively small number of decisions which require the policy evaluation of middle- or upper-managment level decision makers to ensure conformance with program policy and direction.

For evident reasons, our principal focus in this report is on the level of regulatory decisions which require the direct policy involvement of the agency head and the information needed for top- level decision making.

There is another introductory distinction which should be identified. Regulatory requirements may emanate directly from the law, as stated directly by the Congress or as interpreted by the courts, or they may fall within a broader mandate not specifically spelled out either by Congress or the courts but left to the discretion of the agency head to implement. In the case of the former, the law dictates specific standards and requirements which must be administered by the agency pursuant to a specific timetable for compliance. In the case of the latter—discretionary authorities—top-level decision makers are directly involved in broad policy formulation. In these regulatory actions, which tend not to be as well publicized as the specific legislatively dictated programs, many recent regulatory proposals have been preceded by the issuance of an "Advance Notice of Proposed Rule Making," which is designed to ensure greater public participation in the development and specification of the program. We emphasize this because we believe regulatory actions dealing with chemical pollutant control particularly warrant this and other necessary measures to inform and to invite information from the private sector, other public institutions, and the public at large.

DEFINITION OF THE "PROBLEM"

The first step in the initiation of a regulatory action is the perception, identification, and definition of the "problem" to be addressed. By its nature, the emergence of a problem requires a decision. The decision may be to take affirmative action immediately, to postpone action for a fixed period of time, or to postpone action indefinitely, which may be tantamount to taking no action.

Problems requiring a regulatory response generally appear in one of three ways: (1) a "crisis" occurs either as a result of public concern about a perceived hazard or the fortuitous discovery of a potential hazard by an official agency; (2) methodical and systematic review and assessment of known potentially toxic substances confirm the need for action; or (3) legislative requirements or judicial decisions dictate specific regulatory actions.

A "crisis" problem inherently involves a high degree of public awareness and a high expectation that some action will be taken. Because of the public exposure and the concomitant political overtones, there is a compelling need to act, in many cases almost irrespective of the merits and the adequacy of the data and information base. This situation, in our view, presents the decision maker with the most difficult set of circumstances on which to make a judgment about an appropriate regulatory action.

The usual, and probably correct, decision in most crisis situations is to postpone action for a fixed period of time until a preliminary assessment can be made. Obviously, if the known or potential hazard is serious and the population affected is large, the "cost" risk of postponing action may be high and, therefore, action may be warranted even on the basis of relatively little information and data. On the other hand, given the same facts, the decision maker also must evaluate the "cost" risk of making a wrong decision on the basis of too little certainty of the real nature and extent of the problem. If there is some assurance that better, more complete information and data will become available within a relatively short time, a fixed time postponement of action can probably be justified. This would be particularly true if the cost of immediate regulatory action is high. One of the most important considerations for the decision maker is the quality of the data in hand at any given time. A relatively small amount of consistent, reliable data will be a better guide to the timeliness and appropriateness of action than a larger volume of inconclusive or contradictory data.

The optimal circumstances in which to make a judgment concerning the appropriate regulatory action are created by systematic confirmation of the need to control a given chemical. The risks have been identified and evaluated, and the costs are known with some degree of certainty. The benefits of the substance have also been considered and evaluated. The information and data base includes "feedback" from previous actions involving similar or related substances and "feed forward" results of previously planned scientific studies. The substance comes up for regulatory action within a previously assigned priority system which considered the severity of the potential hazard, the size and character of the population at risk, and the viable alternatives for control or mitigation of the hazard. Unfortunately, this best set of circumstances has seldom, if ever, been known to occur. For all of the reasons discussed in this report (and many more not covered in this report, such as deficiencies in manpower and resources), decisions will continue to be made (because they must be) on the basis of less than adequate information and data. We believe some short-term improvements are possible, however, with moderate adjustment in the information- and data-gathering techniques and decision-making processes of EPA and other key regulatory agencies.

The third area of problem definition noted above—specific legislative or judicial direction—also deserves some attention. To the extent that Congress can provide expanded, needed authority to EPA to require more adequate testing of chemicals before they are introduced into the environment, and to prohibit or limit the manufacture, distribution, and use of chemicals which are found to present unreasonable risks to health and the environment, together with increased funding for scientific research, the definition of real and potential problems can be greatly enhanced. Moreover, such expanded authority and resources would enable EPA to move faster in the direction of establishing the kind of information-gathering and decision-making processes discussed in this report.

INFORMATIONAL NEEDS

As noted in the Introduction, we are primarily concerned about the data and information needed to support decision making at the top level of EPA. The most significant regulatory decisions of the Agency pertaining to the control of toxic substances are those involved in the development, review, and approval of the standards, guidelines, regulations, and reports to Congress, and those involved in the case-by-case administrative adjudication of issues elevated to the Administrator for decision.

All other things being equal, decision making in any organization is seldom, if ever, better than the type, amount, and quality of the data and information which flows into and through the decision-making process. It is clear, however, that an adequate data base alone does not insure optimal decision making. The way EPA develops and handles data and information is crucial to the sound, timely resolution of problems presented by toxic substances in the environment. A vast array of information is required relating to the social, economic, scientific, and environmental implications (hazards, costs, benefits) of the use of toxic substances. This section examines several critical aspects of EPA's organization and procedures for developing and transmitting information to the top decision makers and suggests an expanded system for gathering information to enhance the ability of top decision makers to make scientifically sound and socially equitable judgments concerning the regulation of toxic substances.

In a recently distributed internal memorandum (EPA Office of Planning and Management, December 6, 1974), EPA expanded an established coordination and clearance procedure for the development, review, and approval of standards, guidelines, regulations, and reports to Congress (see Figure I-1). The procedure includes four decision points involving the Administrator directly (approval of Advance Notice of Proposed Rule Making; briefing prior to external review; approval of Proposed Rule Making; and approval of Final Rule Making). It also includes the establishment of an intra-agency "working group" comprised of the responsible lead office and other appropriate program and staff office representatives, the preparation of a development plan, periodic review at various stages by the Agency's standing Steering Committee (which includes representatives from every program and staff office at the middle management level), and public meetings and the review of public comments following each stage of the publication process.

We believe the procedures outlined above provide an exemplary system for the development, review, and approval of the most important regulatory items emanating from the Agency. But this procedure is only a limited portion of the total information-gathering and decision-making framework encompassing EPA's responsibilities. Most important to this report, the procedure outlined does not attempt to describe how the information and data base needed to make the indicated decisions flows into and through the process. We do not believe that the EPA memorandum was designed to include information and data flow; we also do not believe that this flow has been specifically detailed in any other Agency management procedure.

We have attempted, therefore, to construct a more complete description of the information-gathering and decision-making process of the Agency as we believe it should be designed. We discuss the process specifically with regard to the regulation of toxic substances, but it has wider applicability to the Agency for other decision purposes.

The essential advantages of the suggested process (see Figure I-2) include the following: (1) it is an ongoing process, providing for continuous review and update, (2) to an extent, it functions independently of specific regulatory items passing through the standards and regulations development, review, and approval procedures, and (3) it incorporates several elements which we believe do not currently exist in the Agency for the systematic preparation and transmittal through the system of (a) hazards, costs, and benefits information, (b) feedback information, and (c) feed-forward information.

We have not attempted to assign specific responsibilities to particular program or staff offices for developing or directing the flow of data and information generated by this expanded system. There must be close coordination between the appropriate program offices and the Agency's Research and Development Office, both in terms of long-range information needs surveys and development plans and in meeting short-term information needs which frequently arise in connection with "crisis" problems and administrative adjudications.

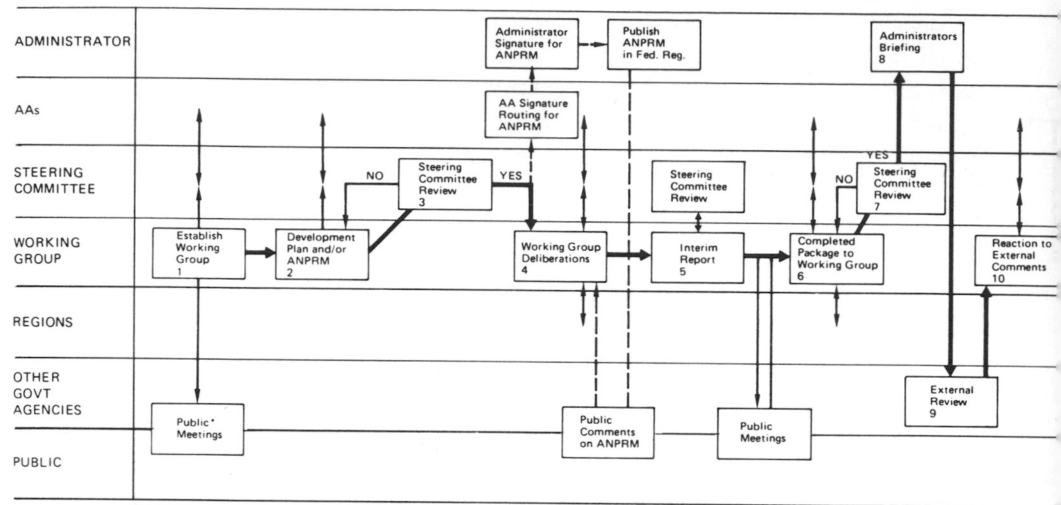

FIGURE I-1 EPA coordination and clearance procedure for the development, review, and approval of standards, guidelines, regulations, and reports to Congress (EPA Office of Planning and Management, December 1974).

HAZARDS, COSTS, AND BENEFITS INFORMATION

Separate reports have been prepared on the assessment of hazards, costs, and benefits associated with the use of toxic substances (see Appendixes F, G, and H). These reports discuss in detail many of the specific informational needs which the Agency should require or provide. We merely emphasize here that EPA's information-gathering techniques should, at the least, include the capability to address the following: (1) identification of the types of hazards presented by a wide range of potentially toxic substances, (2) determination of the minimum data needed to assess the specific hazard potential of a significant number of priority substances, (3) development of a priority system for obtaining specific hazard data on a continuing basis, (4) assignment of responsibility for developing data as between government agencies and the private sector, and (5) establishment of techniques for validating the data generated. In a similar fashion, EPA's information-gathering techniques should include the capability to develop and assess benefits information related to the use of toxic substances.

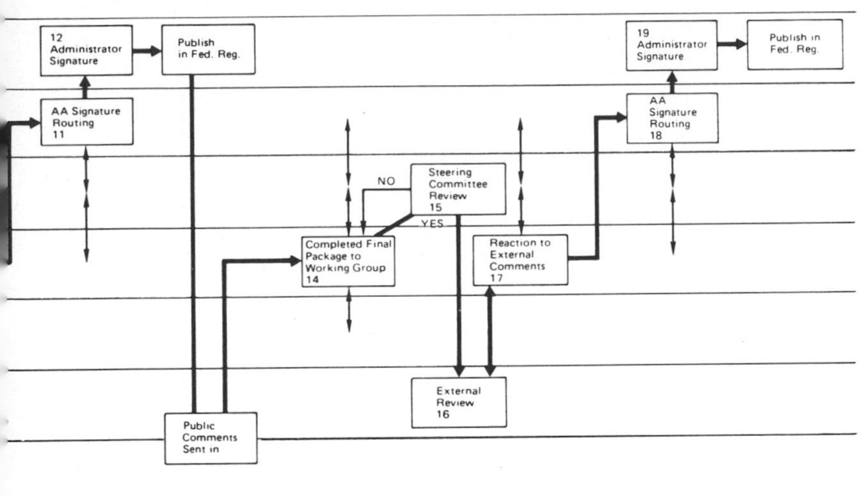

FEEDBACK INFORMATION

Feedback information is that which flows back to the initial decision maker following the execution of a decision and the occurrence of impact. It is distinguishable from "follow-up" action, which the decision maker generally commits himself to at the time of an initial decision in anticipation of future supplementation or revision.

The opportunities for follow-up and feedback information vary considerably with the particular regulatory action involved. Under the Federal Environmental Pesticide Control Act, for example, experimental use permits may be granted in advance of the registration of new products for commercial production. Monitoring conditions included in the permit provide feedback relating to the impact of the initial decision. Follow-up is established in the law, since the applicant may seek registration on the basis of the experimental data obtained.

By contrast, most crisis problems involving high costs of control, political ramifications, and high public visibility, tend to take on overtones of finality once the immediate conflict has been resolved and the issue fades from public attention. These decisions characteristically have not been subjected to adequate feedback information techniques. EPA's auto emission standards, the DDT cancellation decision, predator control poison, and the Reserve Mining case (taconite tailings dumped into Lake Superior) are examples.

In fact, few of these major crisis decisions are final. Administrative

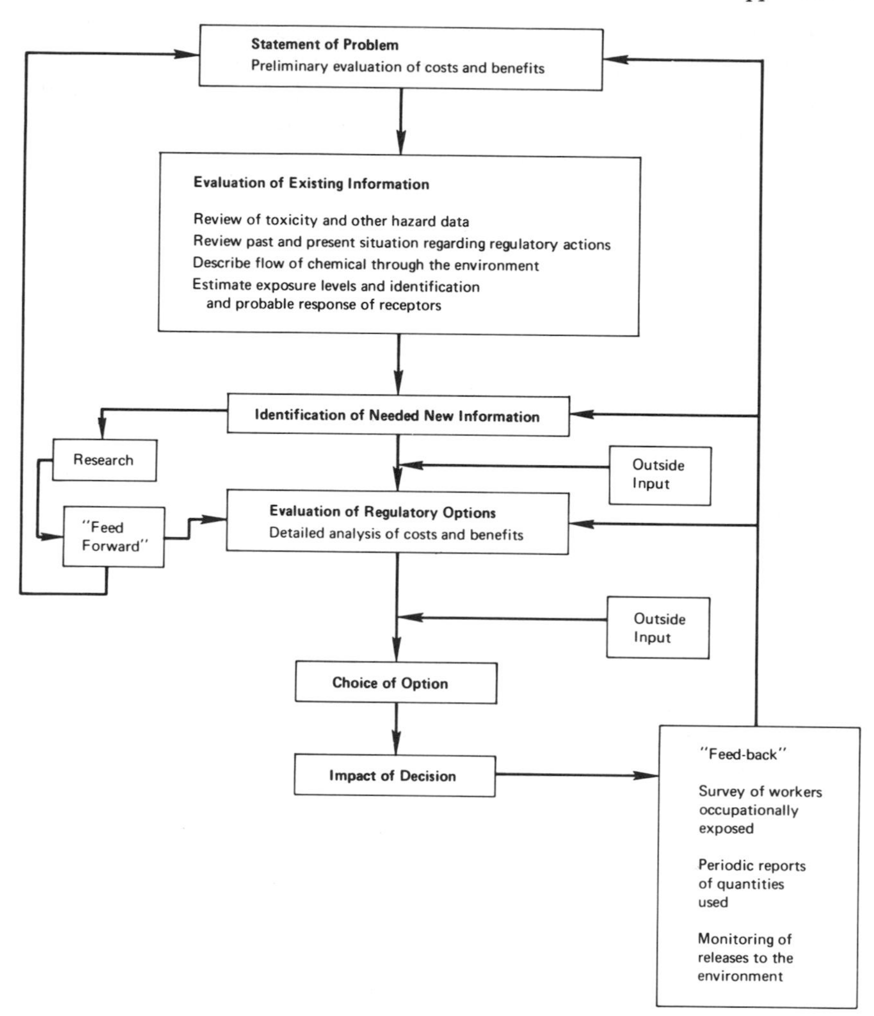

FIGURE I-2 Sequence of steps for nonroutine decision making.

appeals, judicial appeals, and appeals to Congress in these cases are common practice. EPA's regulations, to an extent, help foster a lack of finality about major pesticide decisions by providing for the introduction of significant new evidence following "final" decisions in support of new use registrations, or in support of new cancellations or suspensions. Some limitation is, however, placed on reopening decisions by stringent requirements for determining the significance of new evidence.

In our view, feedback information is especially warranted and necessary in connection with EPA's major regulatory decisions. We believe it is an absolute necessity for the Administrator to display on the public record the information base for decisions in these major cases, even if the effect may be to invest the decision with subjective inferences concerning its "rightness." But gaps or weaknesses in the information and data base should be clearly identified, and steps should be taken to provide for feedback information as well as to spell out in detail the feed-forward information which will be generated by the decision. Moreover, in some cases the Agency should commit itself specifically to perform a formal periodic review (follow-up), particularly where the decision has been to allow a "suspect" product or emission to continue.

In many cases the responsibility for collecting feedback information should be placed on those who benefit from the proposed action. When another government agency (for example, the Forest Service in the tussock moth case) is a proponent, that agency should be responsible for generating feedback information and should also bear the cost of gathering feed-forward information.

FEED-FORWARD INFORMATION

The development and application of a mechanism within the total decision-making process which permits the feed forward of data and information is as essential to optimal decision making as an effective feedback system. As regulatory options are being considered for a given problem, there should be a procedure to ensure that a specific assessment is made of the future information needs associated with each option considered. This serves not only to generate data to feed back to the original decision, but also to feed forward to construct a more complete and validated data base upon which to evaluate future decisions in the same area.

If it is to be useful, the time within which feed-forward data must be generated is of critical importance. The cost of generating feedback and feed-forward data may vary with different options considered. This cost (a transactional cost) must be included in the analysis of the benefit–cost balance for each of the various options under consideration. As noted above, the cost of postponing a specific regulatory action until additional data are generated must also be evaluated.

As noted earlier, another technique is the incremental introduction of a chemical or product in order to obtain more data on its benefit–cost balance, while limiting the extent of the potential hazard to the environment until a more predictable outcome for the various regulatory options can be established. The advantages gained by a limited introduction technique include the continuation of encouragement to innovate with new products and the minimization of capital risk by not permitting full production and distribution (with the attendant economic costs) until a more firmly based decision on whether to allow full production and distribution can be made. The technique also limits the exposure of the population (general and work place) to the particular substance being considered. This technique minimizes the cost of a premature "wrong" decision to allow production where future information might show that the hazards exceed the benefits and necessitate withdrawal of approval, while it encourages introduction of valuable new substances into the marketplace with minimal hazard.

In recent years there have been a few examples within EPA of an optimal feed-forward planning exercise, and those undertaken appear to have been organized essentially on an ad hoc, case-by-case, basis. No effort has been made, to our knowledge, to assess the potential usefulness of a feed-forward system for each regulatory decision.

An Illustrative Case

One recent example of a successful use of the feed-forward planning exercise illustrates its potential usefulness on a wider basis within EPA. When the timetable for sequential improvement in automotive emission controls was established, private industry and federally supported research programs began to search for the most effective technology to meet their standards. Through a series of decisions which are not pertinent here, most U.S. automobile producers decided that, if they were forced to meet the 1975 interim standards, the technology of choice would be the addition of an oxidation catalyst to the exhaust system of conventional internal combustion engines. Preliminary data suggested that hydrocarbon and carbon monoxide emissions would be significantly reduced, while gas mileage would be increased by 8 to 13 percent. By mid-1973, as companies began to tool up for production of the catalysts, some concern began to be voiced within EPA about the introduction of catalysts with respect to both sulfate and platinum/palladium emissions.

After a full review of all available data (one of our panel members felt that the role of possible occupational exposure to sulfates was not adequately covered in either the appropriate legislative mandate or the decision-making process), EPA decided not to forbid the introduction of catalysts in the 1975 model year. The Administrator announced this decision in November 1973, together with a pledge to conduct a high-priority research program to determine whether the concern voiced by critics of the catalysts was valid. If preliminary data supported the concern, the program was to assess the possible technological solutions which could eliminate any substantial human health hazard before the number of cars equipped with catalysts reached a "critical mass" with respect to causing adverse health effects. This time frame was estimated conservatively to be after two model years of catalyst equipped cars were on the road.

Thus, it became essential to define rapidly the areas of agreement and disagreement within EPA's own technical staff, to assess the areas of specific technical expertise and research capabilities within the different offices and laboratories of EPA, to determine what data would be required to answer the questions posed, and to develop a coordinated cross-office/cross-laboratory research program to meet these goals. A task force headed by a representative of the Administrator immediately brought the technical experts from the two "sides" together to define areas of agreement and conflict in a series of three technical papers (Health Effects of Sulfates; Sulfate Emissions from Catalytic Converter Equipped Vehicles; and Projected Human Health Effects from Introduction of the Catalytic Converter), which outlined in detail the areas of conflict (i.e., areas of missing or soft data with wide confidence limits).

Particular emphasis was given to those areas of legitimate concern where data upon which to base a reasonably definite conclusion were lacking. The papers also assessed where, within EPA, the necessary research could be most effectively carried out. Next, the two offices most directly concerned designed supplemental and long-term research programs that would meet the data requirements. Interoffice coordination was increased by establishing a six-man task force (the Assistant Administrator [AA] and a specific coordination/liaison representative from each of the two major offices concerned, the AA for Planning and Evaluation, and a representative from the Administrator's office). The two liaison representatives were given responsibility to ensure optimal communication and cooperation among the different participating laboratories on a day-to-day basis. This was supplemented by regular AA-level briefings, with the AA for Planning and Evaluation and the Administrator's representative arbitrating differences of opinion and informing the Administrator on the progress of the program. This task force was able to expedite the full funding of the relevant research program for fiscal 1975, which established a feed-forward program as having (in this particular case) highest funding priority.

After 12 months, the data generated by this coordinated program have led to an interim issue paper (January 1975), released jointly by the two offices, which substantially supported the concerns about sulfate emissions from catalyst-equipped cars in areas of high traffic conditions. The emission data were slightly modified but were based on much firmer evidence. This issue paper predicted that, after four model years of production of new cars equipped with catalysts, the balance of health benefit from reduction of hydrocarbon and carbon monoxide versus the health disbenefit associated with incremental sulfate emissions would become negative. There are three potential solutions to this problem: power systems which do not use catalysts, gasoline desulfurizaion, or sulfate traps. These will be evaluated by EPA in a time frame that will provide the required data for the next decision point before the fourth model year of catalyst equipped cars starts into production.

This is an excellent example of how a feed-forward system can operate effectively. Obviously, many regulatory decisions do not require such an elaborate, high-level procedure. But a formal mechanism to assess systematically the future informational needs associated with various options under consideration may be useful in many other cases. The needs assessment, including a cost factor, could be included in the deliberations leading to the selection of the most appropriate regulatory option. Where resources are not a limiting factor, there is little reason not to follow this procedure routinely. In most situations resources *are* limited, and following this procedure is a de facto decision that the problem under consideration is of greater significance than at least some other programs from which resources will need to be diverted. Once the option is selected, this should trigger a mechanism to ensure that the planning, support, and performance of the necessary research can be accomplished within an appropriate time frame. For critical programs of highest priority, a system similar to the one described in detail above (including direct participation by Assistant Administrators and the Administrator's Office) would be desirable. For lower priority items, a similar system with a lower level of delegated responsibility could be established.

In this panel's opinion, the identification, generation, and use of both feedback information (to validate past decisions) and feed-forward data (to support more valid future decisions) are of vital importance to the optimal performance of any decision-making system. Formal mechanisms to ensure that these data sources are included in the decision-making process should be developed and institutionalized in EPA.

APPLICATION OF INFORMATION

The organization and application of data and information are a critical step in the total information-gathering and decision-making framework of a regulatory agency. To be useful in the evaluation and selection of regulatory options, the process of organization and application of information must play an integral role in the active "intervention" of the decision-making process. A conscious, coordinated focus within the agency which can coherently organize and array the broadest spectrum of institutional resources is essential. As a first step, it is essential to consider the decision-making process, and in so doing, the organization and array of information at two levels: (1) technically based decisions (often simple "pass/fail" decisions) for the vast majority of specific agency regulatory decisions, and (2) complex, sociopolitical decisions (usually weighing and balancing decisions) for the major regulatory decisions made at the top level of the agency.

In technically based decisions, the function of the data acquisition and organization system is to permit individual experts in technology and science to assess the information gathered and exercise their judgment in a relatively routine pass/fail decision. These technical decisions (on permits, registrations, and so forth) are made in conformance with policy guidelines and regulations approved by higher level decision makers with political accountability. The standards and guidelines provide a sociopolitical consideration to the technically based decisions. At the technical decision-making level, decisions are largely insulated from transient political influence, while still ensuring the application of those broad sociopolitical judgments incorporated in the regulations and guidelines.

Complex sociopolitical decisions, in which major problems facing the agency are decided by the top-level decision makers, include two types of decision which directly or indirectly interface with technical judgments. The first arises when a decision normally made in the technical decision stream is elevated to a higher level because it cannot be resolved by a pass/fail decision, i.e., it is in the "gray" area on the technical spectrum. The other arises when a problem involves mixed technical and sociopolitical considerations and must therefore be elevated to a higher level for decision. In these cases, it is vitally important to define clearly the respective roles of the technical experts and the sociopolitical decision makers. The decision process should require the technical experts to present a full set of options with a corresponding range of benefit–cost–hazard data and explicit statements on the confidence limits of each analysis. The system should minimize the extent to which the technical experts are allowed to superimpose sociopolitical value judgments on the technical data base, since the technical staff serves free of any political accountability.

There are many subtle ways in which individuals without political accountability influence the sociopolitical decision-making process. These influences range from a relatively unintentional bias reflected in the presentation of information to direct advocacy for the adoption of certain preferred options with clear sociopolitical implications. Even more subtle influences, such as those implicit in the selection of participants in a decision meeting, can have a significant impact on the final decision, thus diminishing the degree of freedom of the sociopolitical decision maker in areas where it is important to maximize his flexibility.

In a mixed technical and sociopolitical problem situation, the final balancing must be made by the politically accountable decision makers. It is important to ensure, however, that the final decisions at this level serve the "public good." Safeguards to protect this level of decision making from undue or narrow political pressures should be applied. Political accountability, in itself, is a safeguard, particularly when coupled with free public access to the data upon which the decision was based. In our view this safeguard should be strengthened by publication of a "white paper" describing in detail the rationale used in reaching a given decision. This could be done concurrently with the announcement of the decision, but in some cases the distribution of a decision rationale prior to the completion of the decision process might be advantageous. This could permit constructive criticism as a final test of the "acceptability" of the proposed decision and also serve to reduce the number of decisions which invite almost automatic review in the courts.

A final general comment on the application of information should be made concerning the influence which past policy decisions have on current and future decisions. The agency should conduct a continuous review of the guidelines and/or assumptions on which important regulatory decisions are based, particularly as similar new problems arise for decision. There is a natural tendency to apply nearly identical approaches to apparently similar problems, such as the use of a given mathematical model used to arrive at a risk factor for low-dose exposure to a potential carcinogen when faced with the next problem involving a potential carcinogen. In some cases the same approach is appropriate, but care should be exercised to avoid blind repetition. Without a conscious review of the basic assumptions and methodology, changes which might improve the quality of the technical assessment of the problem can be significantly delayed in their introduction into the decision-making process.

At this stage in the analysis of the information flow, the process of evaluating options for regulatory actions begins to take on new dimensions. Virtually simultaneously, the agency must begin to consider the application of the information available in terms of: (1) a specific focus on the regulatory mechanisms which may be most suitable to the resolution of the problem, (2) a sensitivity to the implications of various regulatory options with respect to other federal agencies, and (3) careful attention to the views of nongovernmental organizations concerning the optional approaches to regulatory action.

PRELIMINARY ASSESSMENT OF REGULATORY OPTIONS

The first question to be asked, given a statement of the problem and an array of available information, is: what, if any, regulatory action is warranted? If the answer is that action should be postponed indefinitely (pending future information or for some other justifiable reason), the appropriate action should be the issuance of a "no action" white paper setting forth the rationale for the decision. If the answer is that action should be postponed for a fixed period of time (pending future information, presumably), then again, among other steps, the decision maker should issue a white paper explaining in detail the reasons for his decision and the steps he intends to take during the period of postponement. As noted previously, this would be an appropriate decision when the data base is inadequate and a delay in instituting some regulatory action would not present a high risk of severe damage to health or the environment.

If the answer is that some affirmative regulatory action should be taken immediately, the decision maker must begin to consider the range of regulatory options available. In this analysis, full consideration should be given not only to the specific regulatory measures provided in the relevant laws and regulations (EPA's and others), but also to other factors, including: (a) the prospective time frame within which the selected option might become effective, (b) informational needs, including an evalution of the costs of instituting various options in relation to their anticipated benefits, (c) the possibility of taking a counteractive action (as opposed to a control action), such as meeting a potential hazard in drinking water by treating the drinking water rather than limiting or prohibiting the use of suspected sources of the hazard, and so on.

INTERAGENCY RELATIONSHIPS

EPA's relationships with other federal agencies concerned with the Agency's regulatory actions should be strengthened substantially. While it is clear that cooperation and exchange of data among interested federal agencies early in the decision process is often useful to the decision maker, there are situations where the early stages of considering regulatory options should be confined within the agency to allow the maximum free flow of ideas and approaches. Interagency access to internal EPA deliberations should begin when the options being considered have been structured in a preliminary form. Early dissemination and cross-fertilization of comments on various options should be structured and not based on free access to exploratory "idea" papers.

There are two goals for interaction among interested federal agencies which must be balanced—cooperation and competition. There are many examples where effective interagency cooperation has enhanced the "public interest" impact of agency decisions. When an important new problem is identified, a great variety of information may be required in a short time. Often it can be obtained more expeditiously and more completely by a coordinated multiagency program, with different parts of the problem delegated to the agency in the best position to complete that task in the time required. Active interagency cooperation is crucial to many agency research programs.

Another consideration is the need to obtain the optimum degree of consistency of approach which different agencies may apply to similar problems. There should be an ongoing procedure for reviewing areas of mutual concern and determining when consistent approaches are the most equitable solution and when divergent approaches are justified. Interagency agreements and memoranda of understanding between agencies should outline explicitly the procedures by which coordination and consistency between the agencies are addressed.

Finally, competition among federal agencies can either hamper or assist in the resolution of problems. Frequently, a diversity of opinion conveyed in frank, open interagency discussions will produce some enlightenment on how the public interest can best be served. This kind of diversity or competition generally is most effective with respect to long-term research and information-gathering activities, but it may also be productive in meeting short-term needs. Constructive interagency competition should not be overlooked as a potential source of information and opinion in choosing among various regulatory options. These competitive systems should serve to provide alternative sources of information to the public and to governmental decision makers.

RELATIONSHIPS WITH NONGOVERNMENTAL ORGANIZATIONS

In the decision-making process, there is a compelling need to provide interested nongovernmental organizations with adequate notification of the regulatory options being considered by an agency. They should be notified early enough in the decision-making process to enable opinions and new data from these sources to exert an impact on the decision process. There should be specific efforts to enlist comments from as broad a spectrum of parties as possible, minimizing the risk of undue influence from well-organized special interest groups through the "Advanced Notice of Proposed Rule Making" system.

Another critical element of nongovernmental input into this system is the role of independent scientific peer review of the technical data base presented to the decision maker. This scientific peer review should be carried out by an institutionalized group with a clearly defined role and responsibilities. This group, chosen by the Administrator after broad formal consultation with appropriate national scientific organizations, should report as an institution directly to the decision maker. The responsibilities of such a "Scientific Advisory Board" must include the review of the technical data base and analysis presented to the decision maker. To accomplish this, the board must have access to the data and deliberations of an agency at any stage of development.

This review function could be triggered on request of the Administrator, by a program office which is in the process of developing and analyzing a given data base, or on the initiative of the board itself. The board must have access to all technical analyses and issues at several formal checkpoints in their development in order to identify possible problem areas for technical review. The board must also have staff and financial support to accomplish their objectives and should be able to call upon other nongovernmental experts for assistance in analyzing specific problem areas, such as toxic substances or health effects. This might be done by means of standing panels and an authority to establish ad hoc panels to assist in short-term analysis of specific problems.

Such a system parallels the recently established Science Advisory Board (SAB) at EPA. Several extensions of the SAB's current authority might be considered. If the SAB felt that the technical analysis in a given situation was inaccurate or misleading (e.g., significantly misjudging the confidence limits of the data) and should be revised before any decision is made, it should report this to the decision maker. If the decision maker then decided to disregard the SAB's recommendation, he should be forced to include an explanation of this decision in the public record. If the SAB feels additional data, controlled by a nongovernmental source, are essential to a valid analysis of a given problem, the Board should have a formal mechanism to request this information from the source. If the data are still withheld (and are not proprietary), the Board should be able to request that the decision maker exercise his authority to subpoena the information. If the decision maker did not respond within

60 days, the Agency Counsel would be instructed to process the subpoena. If he responded negatively, any denial along with its rationale would, again, become a part of the public record.

Scientific advisory boards have a checkered history in performing their presumed function, i.e., serving the public good. They have been especially useful either in improving the technical analysis of the data base or in assisting the decision maker in peer review and acceptance of the validity of a data base and analysis that points inescapably to tough, often politically unpopular decisions. These boards can become counterproductive, however, especially if there is not a clear definition of where review of technical analysis ends and where application of sociopolitical judgments leading to a decision begins. If an advisory board begins to develop advocacy positions regarding the choice of a certain option, it steps past this boundary and attempts to make political decisions without real political accountability. If such a board can perform its task well and avoid attempts to make decisions or co-opt the administrator's role in decision making, while maintaining independence in the exercise of its collective technical judgment, it can serve an extremely useful role in the decision-making process.

AUTHORITY FOR ACTION

The current statutory authorities in EPA for regulating chemicals—the Clean Air Act, the Federal Water Pollution Control Act, the Federal Environmental Pesticide Control Act, and the Marine Protection, Research and Sanctuaries Act (Ocean Dumping Act)—are reviewed in the working paper prepared by the Panel on Governmental Objectives (see Appendix C).

The control processes available under current laws administered by EPA include: (1) labeling and use restrictions on certain pesticides, (2) emissions limitations on air pollutants, (3) effluent standards, limitations, and prohibitions (in certain cases) for pollutants discharged into the waters of the United States or the ocean, and (4) ambient standards for air and water.

Proposed legislation to control chemicals in the environment, such as the pending Toxic Substances Control Act (Senate Bill S.776, February 1975), would expand EPA's regulatory authority to include the following kinds of controls: (1) establishing test protocols for chemical substances which may present an unreasonable risk to health or the environment, to be performed by manufacturers, processors, or importers; (2) premarket screening of new chemical substances; (3) prohibiting the manufacture or distribution of chemical substances, limiting the amount of chemical substances which may be manufactured or distributed, prohibiting the manufacture or distribution of chemical substances for a particular use, or limiting the amount or specifying conditions under which chemical substances may be manufactured or distributed for a particular use; (4) mandating labeling with respect to use or disposal of chemical substances; (5) requiring the keeping and retention of records and the performance of tests to ascertain compliance with other regulatory requirements; and (6) petitioning the courts to restrict the manufacture, processing, or distribution of a chemical substance when the Administrator has reason to believe an imminent hazard exists.

The proposed toxic substances legislation would require the Administrator to consider effects on human health and the environment, the benefits of the chemical substance for given uses, and the availability of less hazardous substances for the same uses in establishing requirements which prohibit or limit the manufacture or distribution, prohibit or limit the use, or otherwise control chemical substances as indicated above.

Chemical substances are used and introduced into the environment in a large variety of ways and may be subject to one or more of the existing controls specified in the laws currently administered by EPA. The theory of the water pollution law is that effluent standards for toxic pollutants would specify minimum control measures which would be required before any toxic pollutant could be discharged into the nation's waters. In fact, the toxic pollutant standards have encountered practically insurmountable difficulty, and there seems to be little prospect of effective regulatory controls stemming from this process in the future. Discharge permits under the water pollution law, however, can achieve substantial control over chemicals proposed to be discharged into the water. Until the enactment of toxic substances legislation or a more workable scheme for regulation under the water law, principal reliance must be placed on the permit system.

The ocean-dumping law also contains flat prohibitions against the disposal of certain dangerous substances (e.g., chemical–biological warfare agents) and restrictions on the disposal of other hazardous or potentially hazardous chemical substances. Last year, the agency refused to permit the disposal of materials containing high concentrations of organic materials (principally ethylene glycol) and antimony compounds in the Gulf of Mexico because of the inadequacy of information on the potential toxicity of the substances contained in the waste.

Each of these potential controls should be evaluated carefully, considering the nature of the regulatory action which may be most suitable to solve the identified problem, the costs and benefits associated with the particular control measures considered, and so on. The optimal control process in some cases, will be that which would intercept a potentially hazardous pollutant before it enters the environment, either through a prohibition or limitation of its use.

Until the last few years there were many gaps in authority for control of chemicals in the environment. If the proposed Toxic Substances Control Act is passed, virtually all the gaps will be closed. One area remaining is air quality in residences, although even in this case, there are some authorities that can be invoked (e.g., control over polluting consumer products may be possible under Consumer Product Safety authorities). In the absence of legal authorities, "jawboning" has been used successfully to persuade industries to resolve the offending problem. In some cases, as for example with polychlorinated biphenyls (PCBs), simply sharing information with the offender may achieve a solution. Authority should be sought from Congress when important gaps become evident.

CHOICE OF ACTION TO ACHIEVE OPTIMUM SOCIAL BENEFIT

There is almost never only one course of action open to a decision maker. One could postulate a mythical case where a chemical pollutant has a precise threshold of effect: above the threshold the effect is catastrophic; below the threshold there is no effect. In this case, there might be reason to decide upon a standard that would offer complete "safety." In reality, however, the effects are dose-dependent, and the costs of control vary markedly with different levels of control. It is precisely this information the decision maker needs so that the implications of the full range of possible options will be evident. Having received the factual information, the decision maker is expected to divine the socially optimal decision. This panel believes there is no way that technicians can help in this process. What is needed is an understanding of how all the "publics" perceive the problem and desire that it be resolved. We can think of no scale or instrument superior to a sensitive and politically responsible mind for weighing these perceptions and desires to achieve a balanced decision. We further note that the care in obtaining the necessary public views should be no less than that taken in securing the "hard facts" that underlie a decision.

As we have noted repeatedly, a decision is not complete until the basis for it (usually including alternatives considered and the reasons for their rejection) has been explicitly set forth in the public record.

Finally, we would note that premarket decisions related to new chemicals probably will require a somewhat different weighing of factors in the decision-making process from that used for existing chemicals already in production. Obviously there will be no *existing* benefits to consider, but if society is to continue to enjoy the positive aspects of chemical innovations, *potential* benefits must be accorded proper weight.

Too often the critical importance of maintaining adequate incentives for innovation is underemphasized. In this context, incremental introduction may prove a useful approach for providing incentive for innovation on the one hand, while minimizing hazard and risk of capital on the other.

APPENDIX J

Working Paper on Market and Private Sector Decision Making

SUMMARY AND RECOMMENDATIONS FOR CONSIDERATION BY THE COMMITTEE

The decision-making process in our private sector, with respect to toxic substances, is complex and unorganized. It is proposed that specific criteria be developed in cooperation with the industrial and scientific communities which will permit a manufacturer to evaluate the relative risks of a new chemical. Management would then have several options with respect to the distribution and labeling of the product. In every instance, federal agencies would have the opportunity to review the appropriateness of the options chosen by management and to force a change to a more stringent risk category if deemed desirable. In addition, the scheme has a unique element with respect to assignment of liability. In the maximum risk category for marketed products, a tax is imposed and held in a trust fund against future claims for damages to persons, livestock, wildlife, or the environment. In the next most dangerous category, represented by products which are cleared prior to marketing, no tax on the product is imposed but liabilities for damages are shared by the manufacturer and the clearing agency. In the other four less dangerous categories, the liabilities are attributable solely to the manufacturers. For each category, product labeling is required which will enable intermediate processors and end users to become familiar with the level of risk and the nature of the product in more precise and meaningful terms than heretofore possible. Thus the labeling acts as a built-in educational mechanism.

It is also recommended that manufacturers be required to report on a timely basis all complaints concerning health and environmental effects of their products, as well as the disposition of those complaints. This is the beginning of a surveillance system which will require substantial expansion to minimize risks in the years ahead.

A CONCEPTUAL PROCESS FOR DECISION MAKING

The findings of this panel are based on fundamental premises concerning the nature of our socioeconomic system. These simple truths are generally shared by all, but they are neglected at times in the search for more elegant decision processes.

1. A free society cannot be maintained without a free market. Political and economic freedoms are inextricably intertwined.
2. Laissez-faire is an anachronism. In today's complex society, where every person's livelihood and security depends not only on his own efforts, but on the efforts of everyone else, the government must set certain rules to assure that the system functions for the benefit of all.
3. The market system will continue to operate with the primary objective of the providers being rewards for capital risked and the primary objective of the users being the satisfaction of needs through the exercise of free choice according to individual desires.
4. The free market concepts can be applied to resources in the public domain—the common goods—through appropriate systems. Many industries use water and air and similar common goods. As objects of value, these goods must be charged for, and the public reimbursed, in proportion to the amount of resource expended. For those resources in limited supply, an allocation basis, as well as cost or replacement charge basis, may be used.

As long as common goods were in unlimited supply, they were free for the taking. Our society considered these goods without commercial value as long as they apparently could not be depleted. Their value has been recognized as they have become scarce.

5. Individuals will take risks so long as they are self-imposed and freely and knowledgeably chosen. Any workable system of control must recognize this attitude.

Our society's first choice among decision processes remains the marketplace. Our culture is still based on the proposition that each individual, by pursuing his own ends, helps society as a whole. It is the duty of government to protect, as far as possible, every member of the society from injustice or oppression at the hand of any other member of it. We also need government to erect and maintain certain public works and public institutions which can never fall within the interest of individuals. This last modification must, however, be applied with care so that it does not displace our fundamental premise of free choice. The quality of marketplace decisions is dependent on the quality of information available to the citizen and the accuracy of the economic bias presented by the direct costs. The first goal of regulatory decision making must be the improvement of this quality and accuracy. This will entail efforts at public education to increase both quantity and quality of information available to the average citizen. It will also require extensive effort to ensure the inclusion of all direct and indirect costs in the price of the goods.

The defects of the marketplace as a democratic decision process are both real and complex. The growing interdependence in our national and global society requires a growing dependence on the political order to make many of our decisions. The "invisible hand" of pure market economics never could cope with the problem of the public institution (common goods). This must fall in the realm of the political process with its inherent conflict, bargaining, and chaos. The political decision, like the marketplace decision, is the result of many individual decisions. These individual decisions are represented by support of interest groups which, through the bargaining process, coalesce into major power blocs. From these, via the legislative process, emerge decisions which represent the will of the majority, modified by the inputs of all active and legitimate minorities. The use of value concepts to determine reimbursement to the public for depletion of common goods will provide a rational basis for resolution of these problems, provided all interests are properly considered.

Common resource control taxes represent a legitimate and adjustable mechanism for applying free market concepts to public goods.

A process which is dependent on conflict and bargaining for resolution of issues will eventually become dependent on the existence of a regulating function. The job of regulators of the conflict is to ensure equality of information and access to the process and right of appeal for the conflicting interest groups. This regulatory role of overseeing the decision process is quite different from the more commonly assumed role of overseeing the execution of a political decision.

The scope of regulatory information gathering and dissemination should include not only scientific and technical data, but also data as to public opinion and choice. Interest groups, whether public interest or business oriented, have widely varying constituencies in size and wealth. Consequently, they present a distorted view of public opinion. The use of data gathered by the growing discipline of opinion research would lend objectivity to the decision process. The support of such data gathering and dissemination should be made part of the analytic role of regulatory agencies.

Responsibility for decisions is borne by those who prohibit the use of known technology as well as by those who encourage its use. All decisions and prohibitions must take this into account. We anticipate that awards for injury may be made by the courts for excessive regulation.

The decision-making process in the private and market sectors of our society has become increasingly complex in the past three decades. A plethora of new chemicals has been developed and, along with it, an increased sensitivity to the interrelationships between these chemicals, man, and his environment. This has, in turn, led to many direct conflicts between society's desire to provide an appropriate measure of public protection and its desire to preserve an efficient production and consumption system.

Resolution of such conflicts has often not proceeded upon an orderly basis; more often than not, some disaster, or near disaster, has forced public action. The result has frequently been the creation of a new governmental agency, or the expansion of authority of an existing agency with regulatory powers more restrictive than necessary for correction of the original defect or defects.

This panel proposes a conceptual process for decision making which will:

1. afford a substantially greater measure of public protection than exists today;
2. require the private sector to make a realistic assessment of the relative risks involved prior to marketing a new product;
3. permit the responsible government agency or agencies to evaluate and review the decisions reached by the private sector;
4. define liabilities more clearly when damage does occur to man, wildlife, or the environment;
5. facilitate the education of the users, processors, transporters, and consumers within the system; and
6. recognize that responsible producers have no desire to impose unduly hazardous products on the environment, and that in cases of extreme or unknown risk, the products may not become available even for selected uses without some limitations to potential liability.

The essential components of the conceptual process are:

a. a set of scientific criteria which will permit categorization of the product;
b. a table of marketing categories;
c. a set of guidelines for labeling products; and
d. a scheme for assignment of liabilities.

The scientific criteria required for categorization of a product (NAS 1975) are the keystone in the conceptual process. It is recognized at the outset that many difficulties will be encountered in the construction and application of a criteria table for use by private industry. Nonetheless, it is felt that the process, which requires the manufacturer to make the initial assessment against these criteria (developed on a cooperative basis between regulatory agencies, industry, the scientific community, and representatives of the general public) in order to select a marketing category, is a significant step toward fulfillment of the objectives of protecting the public interest, preserving the desirable characteristics of our economic system, and fulfilling certain limited needs for hazardous materials.

Further, it is believed that the proposed process will facilitate the assessment of liability in that it will provide the judiciary with a set of scientific criteria for determining relative risks and assigning liabilities.

At this time, it is impossible to project the volume of damage suits that may arise. It is suggested that, initially, the current system of court adjudication be used, and that consideration be given at a later date to establishment of a special arbitration system to facilitate negotiated settlements if the volume of cases warrants it.

The final component is that related to product labeling. Through this mechanism, it is believed that significant gains can be made in educating all parties involved in this complex chain from manufacturer to user.

The sum of the parts of this conceptual process is a system with far-reaching impact. Although it may be imperfect at the outset, it is a system that can be built upon, improved, and made to serve the interests of society.

CRITERIA

Evaluation criteria for chemicals (NAS 1975) are essential to effective functioning of integrated public and private sector decision making. This panel proposed that agencies responsible for the regulation of hazardous chemicals establish criteria involving a scale of values for the following measures, or combinations thereof, to characterize the risk or hazard represented by any chemical substance:

1. bioaccumulation;
2. toxicity properties;
3. persistence;
4. difficulty of detection;
5. dispersion–mobility;
6. degradation products and impurities;
7. index of exposure (volume).

All these measures will not be applicable or pertinent to all substances. Some materials are readily identifiable as having zero values in one or more of these measures. Furthermore, because of the lack of knowledge of the significance of absolute values for bioaccumulation rates and quantities and for toxic properties, the use of well-known compounds as standards for relative risk assessment is proposed. Parallel study of standard substance and test substance should do much to minimize uncertainties in biological assay procedures and should provide a confidence base for test results.

Persistence in the appropriate component of the environment (or in the human body) seems readily quantifiable in terms of half-life of decay or of residence time in an organism exposed to a substance. Even so, use of standards of comparison, particularly in the case of residence time, may be an effective approach.

Difficulty of detection is considered important, especially for those substances hazardous at low concentrations and difficult to identify and quantify.

Dispersion–mobility characteristics are proposed as another effective measure of microscale characterization of health and environmental risk. Here again, partition coefficients, for example, may be productively compared to those of known substances.

Degradation or transformation products and impurities must be considered and assessed by the same criteria. Appropriate study of the chain of physical and biological degradation to naturally occurring substances will be necessary.

Index of exposure is a significant factor in relative risk assessment, for it represents another dimension of dispersion and mobility on the macroscale.

The establishment of criteria by federal agencies should be done with scientific and technical advice from academia and industry, followed by open review and comment sessions for all interested parties. Manufacturers could then decide which category of risk a product falls into, and regulatory agencies, as appropriate, could verify this private sector decision. Users, particularly industrial purchasers of substances as components of another product or for use in manufacturing processes, could request verification by the supplier or by regulating agencies. The early-warning assessment procedure affords the private sector a maximum of decision-making latitude but retains access for governmental intervention and review on an established evaluation basis, thus providing the needed public protection.

MARKETING CATEGORIES

Category I: open marketing with appropriate labeling to be permitted at the option of the producer. No chemical substance should be placed in such a classification if it represents a potential hazard above the prescribed cutoff point on one or more of the standard criteria.

The producer *notifies* the appropriate agency of intent to market a chemical product and includes in such notification data showing properties of the substance with respect to each criteria. This includes all environmental and toxicological data developed by the producer.

No response by the agency need be required, but in cases of valid technical concern based on the data submitted, an agency may require that the substance be classified in a higher category.

Category II: a producer may *register* with the appropriate agency any chemical product intended for marketing, including intended end use. Such registration would require that the producer furnish data defining the degree of hazard for each standard criterion. The agency would acknowledge such registration prior to review of the data, but under its option in cases of valid technical concern based on later review of the data, the agency may require that the substance be classified in a higher category.

For materials in this and all higher categories, the producer should report to the agency within 30 days after receipt all complaints involving health and environmental effects and their disposition.

Category III: a producer may request *certification* of the validity of data which defines the degree of hazard in the intended end use for each standard criteria. The appropriate agency should confirm such data and issue certification, but it may, at its option, require that the substance be subjected to premarket clearance. Such certification may be used in courts, or elsewhere, to establish the good faith efforts of the producer to ensure public safety.

Category IV: for all substances for which a substantial hazard exists as demonstrated by levels above a predefined cutoff point under one or more of the standard criteria, a *premarket clearance* for each intended end use should be required. For such a substance, all the procedures for certification would be required.

In addition, an analysis of economic and social costs and benefits must be made as one factor in rendering a judgment as to the desirability of introducing the product. Such judgment must also include appropriate consideration of the current state of public opinion and social and environmental standards. If such judgment is favorable, marketing for specified end uses with appropriate labeling would be permitted. Any further new use would require a new clearance.

Category V: in those cases where judgment, reached with the help of benefit–cost analysis, is unfavorable, a finding of need for restricted marketing for specified end use or uses may be made. In such cases, no profit-oriented firm will be likely to accept normal product liability responsibility. Therefore, the agency for such cases may institute both a control tax and limited use restrictions. Such controls will prevent the inequitable denial of potential benefit which could result from a total ban, and the control tax would be considered a source of funds for liability which arises through use of the product or lack of knowledge of its effects.

Category VI: a substance may be placed in this category upon determination by the appropriate agency of unacceptable risks associated with distribution and usage. This would ban all distribution of the product.

In those instances in which irreconcilable differences exist as to which category is more appropriate, the matter should be referred to a scientific review panel appointed on an ad hoc basis. The panel's report should be a matter of public record, and its findings should be binding upon the administrator.

As new data become available on any hazard criteria indicating higher values than previously accepted, an agency may recommend recategorization of a product. If an agency proposes such a change, they shall provide the producer(s) with the new data which appear to warrant such a change. In case the data show the substance to have properties which would require premarket clearance if the material were a new product, the procedures for that category shall be followed. In those cases where there exists an imminent hazard to public health, the agency may require immediate reclassification, including the banning of further distribution. Such action must not, however, remove the right of appeal.

LABELING

A chemical substance should be so labeled as to identify the environmental hazard category in which it has been classified. Labels on products in Categories I and II should also indicate the full acceptance of all product liability by the manufacturer. Labels on Category III products should state that the material has been examined as a potential environmental hazard and that the accuracy of the test data is certified. Labels on Category IV products should state that the material is an established environmental hazard and indicate allowed uses. Labels on Category V products should indicate that the material is subject to restricted use, that a control tax has been paid, and that the manufacturer has limited liability.

SURVEILLANCE SYSTEM

It is recognized that the system proposed here does not have a built-in monitoring or surveillance component other than the requirement that the private sector report all complaints involving health or environmental effects and their disposition within 30 days after receipt.

In addition, agencies would have the authority to examine all pertinent company records related to chemicals and products which may have an effect on health or the environment or both.

It is further recommended that federal agencies encourage and support the development of a national surveillance system which would provide early warning of health hazards related to long-term toxicity, including carcinogenicity and mutagenicity.

REFERENCE

National Academy of Sciences (1975) Principles for Evaluating Chemicals in the Environment. Washington, D.C.: National Academy of Sciences.

APPENDIX K Biographical Sketches of Committee Members

J. CLARENCE DAVIES, III, is a research associate at Resources for the Future, Washington, D.C., where he is writing a book on how problems become political issues. He received a B.A. in American Government from Dartmouth College in 1959 and a Ph.D. in Political Science from Columbia University in 1965. Dr. Davies is a specialist on the political and administrative aspects of environmental problems. He was a Senior Staff Member with the Council on Environmental Quality before moving to Resources for the Future. He has worked as Chief Examiner for Environmental and Consumer Protection for the Bureau of the Budget, taught at Princeton University and Bowdoin College, and written two books and numerous articles.

RICHARD FAIRBANKS is a partner in the Washington, D.C., law firm of Ruckelshaus, Beveridge, Fairbanks & Diamond. He received an A.B. from Yale University in 1962 and a J.D. from the Columbia University School of Law in 1969. Mr. Fairbanks served in the federal government as Special Assistant to the Administrator of the U.S. Environmental Protection Agency and subsequently as Associate Director of the Domestic Council in the White House for Natural Resources, Energy and Environment. He has been an adjunct professor of Environmental Law at Georgetown University Law Center and is currently a member of President Ford's Citizen's Advisory Committee on Environmental Quality.

A. MYRICK FREEMAN, III, is an Associate Professor of Economics and Department Chairman at Bowdoin College, Brunswick, Maine. He received an A.B. from Cornell University in 1957, an M.A. in 1964 and a Ph.D. in 1965 from the University of Washington, Seattle. One of his areas of expertise is the economics of environmental quality in which he has published prolifically. Dr. Freeman has also taught at the University of Wisconsin and has been a Visiting Scholar at Resources for the Future, Washington, D.C.

JAMES D. HEAD is currently Manager of the Strategic Planning Consulting Service for the Dow Chemical Company, Midland, Michigan. He received a B.S. in Chemical Technology from Iowa State University in 1943 and a Ph.D. in Organic Chemistry from Columbia University in 1946. Dr. Head has directed overseas plants for Dow and has been Assistant Director of Business Development for Dow Chemical International and Assistant Director of Corporate Planning. He is an expert on the perspective of industry on chemicals in the environment.

JAMES L. GODDARD is Chairman of the Board of Ormont Drug & Chemical Company, Englewood, New Jersey. He received an M.D. from George Washington University in 1949 and an M.P.H. from Harvard in 1955. Dr. Goddard was Assistant Surgeon General, USPHS, Department of Health, Education, and Welfare from 1951 to 1959 and Commissioner of the Food and Drug Administration from 1966 to 1968. He is an expert in the fields of environmental health and public policy.

DAVID L. JACKSON, presently completing a Chief Residency in the Department of Neurology at The Johns Hopkins Hospital, Baltimore, Maryland, will assume the position of Director of the Medical Intensive Care Unit at Case–Western Reserve University Hospitals in Cleveland, Ohio, in September 1975. He received an A.B. in 1961, and a Ph.D. with distinction in Neurophysiology in 1966 from The Johns Hopkins University. He received his M.D. from The Johns Hopkins University School of Medicine in 1968. Dr. Jackson is board certified by the American Board of Internal Medicine. He has served as a Special Assistant to the Administrator of the U.S. Environmental Protection Agency and specializes in the role of government in regulating chemicals in the environment. He also serves as a Councillor to the Section on Environmental Health Sciences of the Pan American Medical Association.

STANLEY LEBERGOTT is a University Professor in the Department of Economics at Wesleyan University, Middletown, Connecticut. He received a B.A. in 1938 and an M.A. in 1939 from the University of Michigan and did postgraduate work at Harvard University in 1958. Professor Lebergott is an expert in the field of economics and public policy. He has worked for the U.S. Bureau of the Budget as an

Economist and as Assistant Division Chief, for the International Labor Office, the U.S. Department of Labor (DOL), and the Institute for Advanced Study. He has served on committees for the National Science Foundation and the Office of Science and Technology and has consulted for the Council of Economic Advisors, DOL.

NORTON NELSON is a Professor of Environmental Medicine, Chairman of the Department, and Director of the Institute of Environmental Medicine at New York University Medical Center. He received an A.B. from Wittenberg College in 1932, a Ph.D. from the University of Cincinnati in 1938, and an honorary D.Sc. from Wittenberg University in 1964. He is a specialist in environmental health, focusing on hazardous and toxic substances. Dr. Nelson has served on numerous advisory committees to various government agencies and private institutions, among them the National Institutes of Health, National Institute of Environmental Health Sciences, the President's Science Advisory Committee, the Office of Science and Technology, the Department of Health, Education, and Welfare, and the Food and Drug Administration.

GLENN PAULSON is the Assistant Commissioner for Science in the New Jersey Department of Environmental Protection. He received a B.A. in Chemistry from Northwestern University in 1963 and a Ph.D. in Environmental Sciences and Ecology from The Rockefeller University in 1971. Dr. Paulson worked as Staff Scientist and then Administrator in the Science Support Program of the Natural Resources Defense Council (NRDC), and prior to that, he worked as half-time Executive Director of the New York Scientists' Committee for Public Information and half-time as a Staff Scientist for the NRDC. Dr. Paulson's expertise lies in the fields of environmental science and public policy.

DAVID P. RALL is Director of the National Institute of Environmental Health Sciences, NIH, Research Triangle Park, North Carolina. He received a B.S. from North Central College, Illinois, in 1946, an M.S. in 1948 and a Ph.D. in 1951 in Pharmacology from Northwestern University, and an M.D. from Northwestern University School of Medicine in 1951. Dr. Rall has extensive experience in comparative pharmacology, cancer chemotherapy, toxicology, and drug research and regulation. He has served in various capacities with the National Cancer Institute and the U.S. Public Health Service.

SHELDON W. SAMUELS is the Health, Safety, and Environment Director for the Industrial Union Department of the AFL–CIO. He was a preceptorial student in philosophy and has done graduate work in theoretical biology at the University of Chicago. Mr. Samuels was Chief of Field Services for the U.S. Environmental Protection

Agency's Air Pollution Control Office, has worked for the New York State Health Department, and is currently on the faculty of the Mount Sinai School of Medicine. He is Vice President of the Society for Occupational and Environmental Health.

JESSE L. STEINFELD is Chief of Medical Service at the Veterans Administration Hospital, Long Beach, California, and Professor of Medicine at the University of California at Irvine. He received a B.S. from the University of Pittsburgh in 1945 and an M.D. from Western Reserve University in 1949. Dr. Steinfeld was Surgeon General of the U.S. Public Health Service from 1968 to 1972, has been a senior investigator at the National Cancer Institute, and has taught medicine at the University of Southern California, Los Angeles. Dr. Steinfeld's expertise lies in the fields of cancer research and oncology as well as in public health policy.

WILLIAM A. THOMAS is a research attorney with the American Bar Foundation, Chicago. He received a B.S. from Purdue University in 1960, a Ph.D. from the University of Minnesota in 1967 with main interests in ecological sciences and natural resource management, and a J.D. from the University of Tennessee in 1972. Dr. Thomas specializes in interactions between law and science. He taught environmental law at the University of Tennessee and conducted research at Oak Ridge National Laboratory.